Sky Is Falling Theories

With a Focus on Anthropogenic Global Warming

Brad Fregger

Groundbreaking Press
Fairfield, Iowa

Sky Is Falling Theories

By Brad Fregger
© Brad Fregger, 2019

Groundbreaking Press
2050 Ruby Lane, Unit 1
Fairfield, IA 52556
512-567-8780
www.groundbreaking.com

First Edition

Senior Editor
Barbara Foley

Book Design & Layout
M. Kevin Ford

Cover Design & Production
M. Kevin Ford

ISBN: 978-1-7339001-4-0

All rights reserved. No part of this book may be reproduced or utilized in any form or by any means, electronic, or mechanical, including photocopying, or recording, or by an information storage and retrieval system, without permission in writing from the author.

Some Quotes to Contemplate as You Read this Book

Thomas Alva Edison

"To raise new questions, new possibilities, to regard old problems from a new angle, requires creative imagination and marks real advance in science."

"Unthinking respect for authority is the greatest enemy of truth."

"Don't listen to the person who has the answers; listen to the person who has the questions."

"No amount of experimentation can ever prove me right; a single experiment can prove me wrong."

"The important thing is not to stop questioning. Curiosity has its own reason for existing."

"Few people are capable of expressing with equanimity opinions which differ from the prejudices of their social environment."

Richard P. Feynman

"For a successful technology, reality must take precedence over public relations, for Nature cannot be fooled."

"It is necessary to look at the results of observation objectively, because you, the experimenter, might like one result better than another."

"It doesn't matter how beautiful your theory is, it doesn't matter how smart you are. If it doesn't agree with experiment, it's wrong."

"The thing that doesn't fit is the thing that's the most interesting: the part that doesn't go according to what you expected."

"The first principle is that you must not fool yourself and you are the easiest person to fool."

"Trying to understand the way nature works involves a most terrible test of human reasoning ability. It involves subtle trickery, beautiful tightropes of logic on which one has to walk in order not to make a mistake in predicting what will happen."

To

All my liberal friends, too numerous to mention

Books by Brad Fregger

Lucky That Way
Stories of Seizing the Moment While Creating the Games Millions Play

Get Things Done – Second Edition
Twelve Secrets of Creating and Leading Exceptional Teams

Get Things Done
Ten Secrets of Creating and Leading Exceptional Teams

One Shovel Full
Telling Stories to Change Beliefs, Attitudes & Perceptions

Why Publish
Making the Right Choices for Your Book

My Thinking Cap
Solutions for Global Crisis

Get Out of the Way!
You'll Never Manage Your Way to Great Leadership

Why Does Anybody Believe in God?
An Essay on Creation

Venturing Beyond Earth
The Need for Exo-Terra Sustainable Cultures

The Sky's the Limit
How to Effectively Market Yourself

In Production

A Surprising Life
Stories from My Journey

Booklets

Publish!
Maharishi Mahesh Yogi and Conservative Thought

Preface

It seems that Sky Is Falling theories have existed for as long as we have been on this planet. Whenever human beings experienced anything unusual, whether it be a devastating weather phenomena, a solar eclipse, or a violent thunderstorm; there were always "priests" screaming that the people had angered the gods and that they, the priests, were the people's only hope for redemption. Only they could appease the gods and return peace and tranquility. There were many ways of appeasing the gods, from rituals and offerings to human sacrifice.

This attitude of "appeasing the gods" was also seen when organized religion took the lead and Giordano Bruno was burned[1] at the stake in February 1600 for daring to go against the scientific consensus that Earth was at the center of the Universe. For most of the Inquisition, which lasted 700 years, "science" was aligned with the church. Bruno wasn't the only one burned at the stake; it is estimated that tens of thousands suffered in this way, which was only 2 percent of those punished by the Inquisition.

Today the priests and religious leaders have been replaced by "scientists" whose inquisitional beliefs and actions are best reflected in their fanatical support for Anthropogenic Global Warming (AGW)[1]. These "scientists" are either convinced that humanity is destroying the planet or maliciously using the public's ignorant concern to advance their reputations and funding.

---

[1] Human-caused catastrophic global warming

They gladly destroy the lives and reputations of those who dare to suggest that the processes that drive Climate Change are not well understood, and that, in truth, current research suggests that the danger is nowhere near catastrophic.

Destroying the integrity, careers, and funding of scientists focused on doing the research necessary to discover the truth is, in effect, just as bad as burning them at the stake. For a scientist who has spent his whole life in the pursuit of truth, this is a fate worse than death. It even gets worse; some idiots actually want opposing scientists, those that do not agree that AGW is a fact, to be arrested and imprisoned for up to 20 years.[ii]

Science always takes a hit when research suggests that something has been factually verified and then the public finds out later that the exact opposite is true. Sky Is Falling myths and theories that gain the public's attention exacerbate the problem; in regard to AGW, we literally have a perfect storm situation. A major portion of the public, the media, governments, Hollywood, and most scientific institutions have accepted the AGW hypothesis as a fact, not the failed hypothesis that it actually is.

But, I'm getting ahead of myself. There's much more to cover before I bring you up to date on the AGW fraud; the biggest hoax ever played on the human race.

Brad Fregger
June 2019

Contents

PREFACE V
CONTENTS VII
TABLE OF FIGURES VIII
INTRODUCTION 1
SKY IS FALLING HISTORY   1 5
   SIXTEEN FAILED PREDICTIONS OF CATASTROPHIC EVENTS 10 ACCURACY OF THE PREDICTIONS OF CATASTROPHE 13
AGW HISTORY   2 19
   SEVEN CRITICAL QUESTIONS THAT MUST BE ASKED 20
   A SHORT HISTORY OF THE AGW HYPOTHESIS 21
PERFECT STORM "SCIENTIFIC CONSENSUS"   3
.... 29 THE OPPOSING VIEW   4
.... 39
   OPPOSING RESEARCH REGARDING SIGNIFICANT AGW ISSUES 42
UNINTENDED CONSEQUENCES   5 55
CONCLUSION 67
ADDENDUM 75
   THE SIXTH MASS EXTINCTION? 75
AUTHOR'S BIO 89
CONTACT 91

Table of Figures

1: The Population Bomb	2
2: Hale-Bopp with Spaceship Following	3
3: Rachel Carson	5
4: Earth Globe	6
5: New York Times, April 22, 1970	7
6: Life Magazine January 30, 1970	11
7: Paul Ehrlich	11
8: Kenneth Watt	12
9: The Sun	13
10: Suicide Rates per 100,000	16
11: Annual CO2 Emissions from Fossil Fuels	22
12: Michael Mann	23
13: Mann Hockey Stick Graph	23
14: Clovis Arrowhead	25
15: Eugene Shoemaker	26
16: Comet Shoemaker Levy Hitting Jupiter	27
17: Arctic Sea Ice Measurements	46
18: Brad Fregger	89

Introduction

"Control is the enemy of evolution."
Maharishi Mahesh Yogi

This book will take a look at the theories that suggest that the world is coming to an end and that human beings (especially western societies) are the cause. I will also cover the potential dangers that these theories (myths) represent to our society and the world in general. There will be a focus on the Anthropogenic Global Warming (AGW) hypothesis.

In almost every case these theories, from the "population bomb" to "human-caused, catastrophic global warming" (AGW) are developed initially by scientists who had noticed an issue that begs research to determine its cause and the potential impact on humanity's societies or Earth in general.

However, the more spectacular theories, the ones that suggest the destruction of civilization and, potentially, the end of life on Earth, attract those "scientists" hungry for credibility and the funding that will enable them to live the lifestyle that they believe is their due.

This includes: funding for their questionable research, a very comfortable living situation, royalty advances for the fear-generating books they write, and high-paid speaking engagements. In addition, they enjoy the adoration of followers who are either intellectually challenged or see these "scientists" as an aid to achieving their own agendas. I will describe this latter group of followers more completely later in this book.

was introduced to the Sky Is Falling community of scientists with the publishing of *The Population Bomb* by Stanford University professor, Paul R. Ehrlich in 1968. In the book Ehrlich postulated that humans would suffer mass starvation in the 1970s and 80s due to overpopulation. As a technological optimist it seemed asinine to me that humanity would allow this to happen.

Figure 1:
The Population Bomb

However, his book, which sold over two million copies, supported the belief that humanity was the problem and that we would suffer the consequences of our actions against Mother Earth[iii], if we didn't change our ways.

Ultimately the entire concept was proven wrong; the surprise is how many were taken in by this idiocy. This is a problem in itself, as the image of the future he is presenting is one that few would like to experience. As a result of such theories, many lose hope, faith in the future, and the society and individuals within it suffer.[2]

Ehrlich's predictions were not only totally wrong, but we are literally feeding the world better today than we were in 1968. For example, there was a 42 percent decrease in world hunger[iv] between 1990 and 2014.

Regardless, Ehrlich still stands by his prediction of the collapse of civilization due to overpopulation, basically stating that he "just had the timing wrong." And there are plenty of people who still support him and his failed prophecies.

---

[2] I cover the impact of images of the future, both positive and negative, in my book, *Society's Image of the Future and Socioeconomic Health*.

It seems that if you are a scientist who is after the actual processes that drive climate change, you're reputation, credibility, and funding can be stripped from you; you can be "burned at the stake." But, if you're a fraud whose research has proven to be a total disaster, you can hold on to both your reputation and funding. An excellent example of the crazy world we live in.

This reminds me of Marshall Applewhite' (one of many who have predicted the end of Earth).

Figure 2: Hale-Bopp with Spaceship Following

Applewhite, leader of the Heaven's Gate cult, claimed that a spacecraft was trailing the Comet Hale-Bopp and argued that suicide was 'the only way to evacuate this Earth' so that the cult members' souls could board the supposed craft and be taken to another "level of existence above human." Applewhite and 38 of his followers committed mass suicide. (There's no way for us to know for sure whether they made it or not.)

In the same way, those that follow Sky Is Falling false prophets like Ehrlich are not only committing intellectual suicide but they are also putting the world in danger. Fear of a human-caused, worldwide catastrophe does no one any good and leads others to take actions that could be extremely detrimental to the entire human race.

Sky Is Falling History   1

I would rather have questions that can't be answered than answers that can't be questioned.

Dr. Richard Feynman

The world was first introduced to the "scientific" concept that humanity could cause great harm to Earth in 1962 with the publication of Rachel Carson's book, *Silent Spring*. Her book stayed on the bestseller list for 31 months and has since sold millions of copies. In addition, in 2006 her book was listed as one of the 25 best science books of all time by *Discovery Magazine*. She is given credit by many as being the inspiration for the creation of the Environmental Protection Agency (EPA).

Figure 3: Rachel Carson

The book represented a watershed moment, the catalyst that raised public awareness and concern for the environment and the dangers of runaway pollution.

Although she remains one of the heroes of the environmentalist movement for her impact on the public's consciousness (or unconsciousness), her major achievement was making DDT the villain that was destroying animal life around the world. This, in spite of The National Academy of Science[vi] concluding in 1965 (3 years after the publication of her book) that DDT had prevented 500 million deaths in the previous 20 years.

The final result is that over 50 million people died unnecessarily in mostly poor tropical countries because DDT was not there to eliminate the mosquitoes that carried malaria and the louse which carries typhus.

In addition, Carson, who had no scientific training, was completely wrong; DDT has a very low toxicity to animal life.[vii]

Then, in 1966 we got our first view of Earth as a planet residing in the immensity of space when the Apollo astronauts sent back the first photos of the globe of Earth. This view of the world was exactly what was needed to convince many of the importance of taking our responsibilities toward the planet/environment seriously.

Figure 4: Earth Globe

This was a *good* thing. The problem is that Sky Is Falling theories continually take us off focus and we are often "forced" to concentrate on issues that end up having little to do with the real problems that we are, or will be, facing.

This great need to be active in saving Earth from humanity's lack of concern for the environment, a need that was supported by the success of both *Silent Spring* and *The Population Bomb*, as well as the Apollo astronauts photos of the globe of Earth, was solidified by the creation of Earth Day, held for the very first time in 1970 and still being held every April 22, (celebrated in over 193 countries).

Figure 5: New York Times, April 22, 1970

The success of Earth Day gave affirmation to those scientists who believed the world was in grave danger. Barry Commoner, considered by many to be the father of the modern ecological movement, said it this way, "Earth Day 1970 was irrefutable evidence that the American people understood the environmental threat and wanted action to resolve it."

I'd like to take a little side trip here and discuss Barry Commoner's Four Laws of Ecology. In addition to clearly defining how he believed we must move forward to protect Earth, I believe they are still extremely relevant. Here are his four laws:

1. *Everything is connected to everything else.*
You cannot change one thing without affecting another. This is well understood by computer programmers who know only too well that making one change to their program is bound to change something else, and that they can't determine where that change will happen or the effect it will have on their program.

2. *Everything must go somewhere.*
Another great truth; when the laws of nature are followed this rule, in many ways, governs the way our planet works. Nature knows best how to recycle, our job is just to make sure that we don't screw things up.

3. *Nature knows best.*
So very true; Maharishi Mahesh Yogi says it like this, "Control is the enemy of evolution." Most of the time, humanity's efforts to improve on nature go astray and we get a major hit by the "unintended consequences."

4. *There is no such thing as a free lunch.*
Ultimately there's a price for everything and many times the price is beyond what we are willing or able to pay.

While I'm sure some current environmentalists would interpret these laws quite differently from my interpretation, I can clearly state that I am in full support of each of the four laws.

The first and second laws are definitely true and must be clearly understood by anybody planning to build a sustainable community. There are obviously ways that we can use these laws to stimulate creativity and innovation; that is, find ways to make them work for us rather than treat them as a barrier to be overcome.

The third law is one that I have been writing about for, literally, decades. We often get into trouble when we do not take the time to gain the needed understanding of the issue. In my management classes I tell my students that you cannot solve a problem, resolve a conflict, or make an intelligent decision without first gaining understanding.[3]

There may come a time when our understanding of how the Universe works will allow us to "play around" in this way; we aren't there yet, in fact our knowledge of how the Universe works is no greater that an infant's knowledge as to how their daddy's car works.

Yes, nature does know best ... no matter how smart our "leaders" think they are.

The fourth rule, tells us that there will be a price to pay. The question, of course, "Will the cost be more than we are willing or can afford to pay?"

For example, let's say that environmentalists had determined that an owl's habitat in the Oregon forest was endangered due to logging. The owl could be saved by setting aside 6 million acres of ancient forest where logging wasn't allowed and then making sure that the habitat was conserved.

However, this decision may destroy the area's logging industry, forcing loggers and supporting services out of business, essentially putting approximately 100,000 people out of work.

In addition, the policies designed to save the forest for the owls may, in fact, actually contribute to the owls decline. And, do we know for sure that the logging of the forest is causing the owls' potential extinction; there may be other forces at work.

There is always a price to pay; this is the bottom line.

*****

---

[3] The degree of understanding needed depends on the complexity of the issue.

With the tremendous interest in how humanity was destroying the planet, the 70s spawned many other predictions by pseudo-scientists and others who jumped on the funding bandwagon, trying to make a name for themselves in the environmental movement by predicting even more catastrophic events caused by human activity. This resulted in a slew of failed catastrophic predictions.

Sixteen Failed Predictions of Catastrophic Events[viii]

1. Harvard biologist George Wald in 1970 estimated that "civilization will end within 15 or 30 years unless immediate action is taken against problems facing mankind."
2. "We are in an environmental crisis which threatens the survival of this nation, and of the world as a suitable place of human habitation," wrote Washington University biologist Barry Commoner in Earth Day issue of the scholarly journal *Environment*.
3. The day after the first Earth Day, the *New York Times* editorial page warned, "Man must stop pollution and conserve his resources, not merely to enhance existence but to save the race from intolerable deterioration and possible extinction."
4. "Population will inevitably and completely outstrip whatever small increases in food supplies we make," Paul Ehrlich confidently declared in the April 1970 *Mademoiselle*. "The death rate will increase until at least 100-200 million people per year will be starving to death during the next ten years."
5. Ehrlich sketched out his most alarmist scenario for the 1970 Earth Day issue of *The Progressive*, assuring readers that between 1980 and 1989, some 4 billion people, including 65 million Americans, would perish in the "Great Die-Off."

6. "It is already too late to avoid mass starvation," declared Denis Hayes, the chief organizer for Earth Day, in the spring 1970 issue of *The Living Wilderness*.
7. In January 1970, *Life* reported, "Scientists have solid experimental and theoretical evidence to support...the following predictions: In a decade, urban dwellers will have to wear gas masks to survive air pollution ... by 1985 air pollution will have reduced the amount of sunlight reaching earth by one half...."

Figure 6: Life Magazine January 30, 1970

8. Peter Gunter, a North Texas State University professor, wrote in 1970, "... by 1975 widespread famines will begin in India; these will spread by 1990 to include all of India, Pakistan, China and the Near East, Africa. By the year 2000, or conceivably sooner, South and Central America will exist under famine conditions. By the year 2000, 30 years from now, the entire world, with the exception of Western Europe, North America, and Australia, will be in famine."
9. Ecologist Kenneth Watt told *Time* that, "At the present rate of nitrogen buildup, it's only a matter of time before light will be filtered out of the atmosphere and none of our land will be usable."
10. Barry Commoner predicted that decaying organic pollutants would use up all of the oxygen in America's rivers, causing freshwater fish to suffocate.
11. Paul Ehrlich predicted in 1970 that "air pollution...is certainly going to take hundreds of thousands of lives in the next few years alone." Ehrlich sketched a scenario in which 200,000 Americans would die in 1973 during "smog disasters" in New York and Los Angeles.

Figure 7: Paul Ehrlich

12. Paul Ehrlich warned in the May 1970 issue of *Audubon* that DDT and other chlorinated hydrocarbons "may have substantially reduced the life expectancy of people born since 1945." Ehrlich warned that Americans born since 1946…now had a life expectancy of only 49 years, and he predicted that if current patterns continued this expectancy would reach 42 years by 1980, when it might level out.
13. Kenneth Watt warned about a pending Ice Age in a speech. "The world has been chilling sharply for about twenty years. If present trends continue, the world will be about four degrees colder for the global mean temperature in 1990, but eleven degrees colder in the year 2000. This is about twice what it would take to put us into an ice age."

Figure 8: Kenneth Watt

14. Ecologist Kenneth Watt declared, "By the year 2000, if present trends continue, we will be using up crude oil at such a rate…that there won't be any more crude oil. You'll drive up to the pump and say, `Fill 'er up, buddy,' and he'll say, `I am very sorry, there isn't any.'"
15. Harrison Brown, a scientist at the National Academy of Sciences, published a chart in *Scientific American* that looked at metal reserves and estimated that humanity would totally run out of copper shortly after 2000. Lead, zinc, tin, gold, and silver would be gone before 1990.
16. Paul Ehrlich predicted that "since more than nine-tenths of the original tropical rainforests will be removed in most areas within the next 30 years or so, it is expected that half of the organisms in these areas will vanish with it."

So, how accurate were these Sky Is Falling predictions? There were four major predictions: massive starvation; air pollution will kill hundreds of thousands; crude oil will be gone by the year 2000; there will be another ice age, also by the year 2000.

Accuracy of the Predictions of Catastrophe

Massive Starvation. While the world still has a problem with millions still undernourished, we have made massive strides and there is no reason this trend won't continue. However, there are two main issues that could easily cause us to lose ground:

1. Current research on the Sun suggests that we may be in for another mini ice age. The relationship between a "quiet Sun" (an absence of sun spots) and global cooling is very strong, and as we learn more about the Sun's life cycles the hypothesis for a drop in the temperature of the planet is becoming more likely. This would, of course, limit the growing cycle and growing enough food for the planet could become more difficult.

   Figure 9: The Sun

2. Ethanol, as an alternative energy source, is not a good idea; we've still got millions of people in the world that are undernourished. Growing corn as an energy source not only uses up an outstanding food source, but it increases the price of food corn to the point where many in third world countries find it extremely difficult to afford to purchase what they need to survive. Since 2006 food prices in some of the world's developing countries have risen 50-75 percent.[ix]

   In addition, ethanol is not easily transported. It can't be shipped in pipelines and must be trucked to where it is needed. In fact, the total process of creating ethanol and delivering it to the consumer is quite possibly a greater problem for the environment than fossil fuels.

Air Pollution: In the United States air pollution is really no longer a problem. However, it is a big problem in China and India. Regardless, CO_2 is not the problem in either of these countries and forcing them to focus on eliminating CO_2 "pollution" is taking them off focus from the real problem, which is the type of air pollution that is an immediate danger to the physical health of their citizens.

The air pollution in China has numerous causes from rapid economic growth which has been a boon for automobile sales, producing massive amounts of pollutants. In addition 70 percent of China's energy comes from coal; burning 3.2 billion tons of coal annually is another major contributor to China's critical air pollution problem.

If these issues were handled, China's air pollution problem would be solved, without touching the additional production of CO_2, which is not a pollutant.

America burns around 1 billion tons of coal; however, our use of coal has been rapidly decreasing from 39 percent of our energy consumption in 2014 to 27.4 percent in 2018. This decrease is mostly due to the increased use of shale gas.[x]

We Are Running Out of Crude Oil: This was supposed to happen by the year 2000, another phenomenal miscalculation. With the discovery of new oil fields around the world and the use of new fracking technologies, it is currently estimated that the world has between 50 and 100 years of fossil fuels left, plenty to last us until effective, efficient alternative energy sources are developed.[4]

A New Ice Age: This prediction was actually turned on its head with the almost total acceptance of the AGW hypothesis. Later in this book I discuss the very real potential that we, indeed, may experience another mini ice age within the next few decades.

Dramatic Decrease in Life Expectancy: In 1970, Ehrlich warned that Americans born since 1946 had a life expectancy of only 49 years.

Interestingly, while Ehrlich was telling us that the life expectancy in 1970 was 49 years, the University of California in Berkeley was given a very different figure 67.1 (men), 74.7 (women).[xi]

However, the most current figures for the United States (2018) put the life expectancy at 78.7 years. Could this pseudo-scientist have been much further off? The life expectancy in other developed countries is even slightly higher, about 80 years.

America's problem is suicide and drugs play a major role in the decrease in life expectancy. In the last 17 years the suicide rates in the United States have risen 25% with some states having an increase of up to 40%.[xii]

Figure 10 shows the suicide rates by state for 2017. It is very interesting that the states showing the most suicides/100,000 tend to be in the west/central part of the United States, flyover country, while California and New York are below the national average.

---

[4] I don't think we will ever run out of fossil fuels; new technologies will develop additional energy sources and ways of storing the energy and this will lower the need for fossil fuels.

I suspect this increase is due, in large part, to the negative image of the future being presented by progressives, the media, and the entertainment industry.[5] Why some states seem to have escaped this scenario is a puzzle.

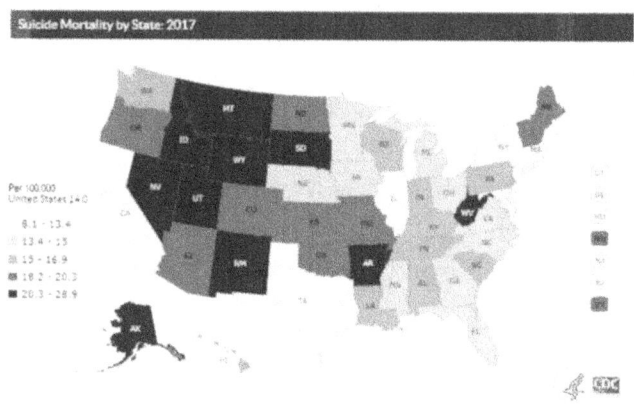

Figure 10: Suicide Rates per 100,000

Take those two issues out of the equation and our life expectancy revels that of any other country. Regardless, we still have a life expectance that is almost twice what the doomsayers predicted.[6]

So, you can see that these Sky Is Falling predictions tend to be totally wrong; they completely rule out Earth's ability to correct itself and humanity's exceptional ability to adapt in ways that

---

[5] I discuss the power of images of the future, both positive and negative in my book, *Society's Image of the Future and Socioeconomic Health*.

[6] Time and time again these pseudo-scientists are proven wrong. What surprises me the most is how intelligent, caring people can continue to give credence to these idiots. There must be a psychological explanation. Perhaps liberals in America feel subconsciously guilty living in a country with such a high standard of living compared to the rest of the world and, therefore, embrace these predictions of doom.

often improve the situation. When we consider that we are about to become a spacefaring race[7], the resources that are available in the solar system should give us easily another 1000 years of prosperity. After that, who knows?

---

[7] My book, *Venturing Beyond Earth*, presents a strong case for humanity becoming a spacefaring race.

AGW History 2

AGW appears the result of political dogma corrupting proper scientific thought. Convenient misinterpretations of basic science in combination with large amounts of confirmation bias have obscured the scientific method. The scientific foundations of the "greenhouse effect" and "radiative forcing" appear baseless.

Dr. Judith Curry

It's time to take a look at the history of the most current Sky Is Falling hypothesis, AGW or human-caused catastrophic global warming.

But before I begin, I want to make the definition of AGW, as it is being used by global-warming fanatics, very clear. They are talking about a global-warming scenario that will be catastrophic to all life on this planet.

There is a strong need, by those who support and depend upon the AGW hypothesis, to convince the public at large of the truth of this *monsters under the bed* myth. In addition, the decades-long pause in warming that Earth has just experienced, forced them to change their terminology and call it "climate change."

The change was only made to manipulate the public into believing there was a true scientific consensus regarding the AGW hypothesis.

Here are a few facts that you should be aware of:

- Every[8] scientist and realist[9] in the world believes in climate change.
- Every scientist and realist in the world believes that Climate Change is a natural process.
- Every scientist and realist in the world believes that Earth is currently warming.
- Every scientist and realist in the world believes that humanity is contributing to the warming of the planet.

Anyone telling you, or even suggesting, that realists do not accept the above as facts, is manipulating you and you should not believe another word that comes out of their lying mouths.

The questions asked, by those who believe that the AGW nightmare scenario has no basis in reality, are very reasonable and need to be asked. As Feynman said, *"I'd rather have questions that can't be answered than answers that can't be questioned."*

Seven Critical Questions that Must be Asked

1) Has the AGW hypothesis accurately predicted warming trends for the past 30 years?
2) Is it possible there are other factors that have an impact on Climate Change that the current models are ignoring?
3) Is the cure worse than the disease? Is there a reasonable possibility that the proposed efforts to "solve" AGW will cause

---

[8] A slight exaggeration, creationist scientists may disagree with some of the issues.
[9] We believe that the designation of "realist" is more accurate than "denier." While we do *deny* the accuracy of the predictions being sold by those who support AGW, our approach is realistic and based on traditional scientific principles.

significantly more suffering than the level of global warming Earth is likely to experience?
4) Is it possible that the increase of CO_2 in the atmosphere is actually a boon to the planet's ecological system?
5) Is the pervasive focus on AGW stealing efforts, resources, and funds from much more worthy societal issues?
6) Are the efforts to solve the problem currently causing unwarranted suffering and death in the poorer nations of Earth?
7) Finally, should the AGW hypothesis be proven to be completely wrong, how much would scientific credibility and the world's economy be harmed?

These are very real issues that must be faced if we are ever to do what's right, make intelligent decisions, and create a planet where nature and humanity live in harmony.

A Short History of the AGW Hypothesis

A Swedish scientist, Svante Arrheniu, was the first to claim that fossil fuel combustion may eventually result in enhanced global warming.[xiii] In 1896 he proposed a relationship between atmospheric CO_2 concentrations and temperature. His research did not have much of an impact because scientists at that time believed that the human influence on climate was insignificant compared to nature's powerful impact.

The concept of CO_2's significant impact on Earth's temperature really didn't start to gain traction until the slight cooling trend of the 1940s to 1970s ended and the temperature began to rise once more. In fact, Earth's mean temperature rose so fast from 1980 through 1998 many scientists and others began to take the possibility of an increase in global warming very seriously. This resulted in the AGW hypothesis that postulated a direct relationship between global warming and humanity's use of fossil fuels increasing the amount of CO_2 in the atmosphere.

This hypothesis initially seemed to be very likely. Earth was warming and we were putting tons of CO_2 into the atmosphere with the pervasive use of fossil fuels around the world. Since CO_2 was a greenhouse gas, it would tend to cause a certain amount of warming.

Figure 12 shows the amount of CO_2 the major contributors have been putting into the atmosphere since 1960. The annual US contribution lowered just prior to 2010 and it has remained basically constant since then.

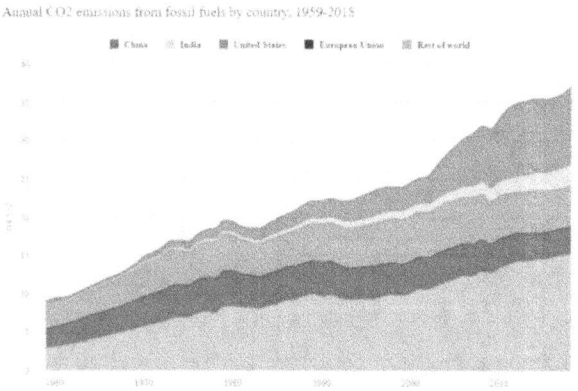

Figure 11: Annual CO2 Emissions from Fossil Fuels

What's critical to this discussion is that the amount of CO_2 we were putting into the atmosphere increased from 370 ppm (parts per million) in the year 2000 to 390 ppm in 2010. Yet, during that same time the world's temperature was relatively flat.

In 1998, Michael Mann, Raymond Bradley and Malcolm Hughes created the "hockey stick graph" model that strongly supported the hypothesis that increases in CO_2 had a major influence on global temperature. In fact, if their model was correct, which showed that if we continued to pump CO_2 into the atmosphere at the current rate, it wouldn't be too long before Earth would become uninhabitable.

Figure 12: Michael Mann

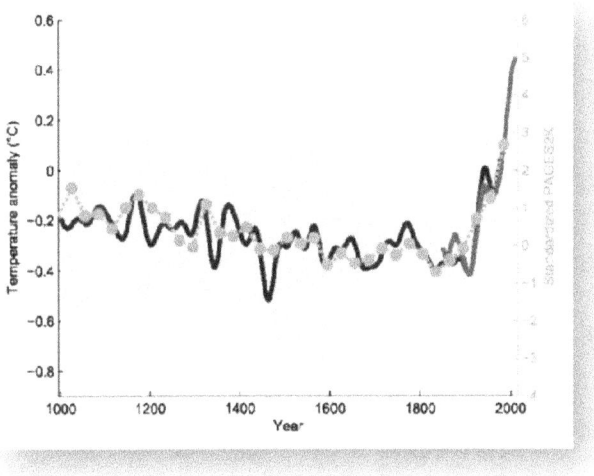

Figure 13: Mann Hockey Stick Graph[xiv]

Models from other climate scientists tended to show the same result and the AGW hypothesis gained significantly new strength, especially in popular culture.

In other words, a "scientific consensus" was born.

Short History of Failed "Scientific Consensus"

It should be noted that the existence of failed "scientific consensus" is not all that unusual. It usually happens when an idea takes an area of the scientific community by storm and everybody buys into it without much questioning.
However, most (if not all) "scientific consensus" is ultimately proven wrong. This is the objective of scientific research, to discover the deeper truth of our Universe or as Einstein stated it,

> *I want to know how God created this world. I am not interested in this or that phenomenon, in the spectrum of this or that element. I want to know his thoughts. The rest are details.*

Earth Centric (Geocentric) Model

The geocentric model created by the Hellenistic astronomer Claudius Ptolemaeus in the 2nd Century became the standard and was the longest living and most devastating "scientific consensus" in history. The Inquisition would not allow anyone to discuss alternative views on a wide variety of issues and all papers or books suggesting that Earth was not the center of the Universe were banned.

Around 1514 Copernicus formulated his Heliocentric hypothesis[xv], which stated, among other things, that Earth was not the center of the Universe. His book describing the Heliocentric astronomical model of Earth and the Planets was published around 1542 but it was ultimately banned. Copernicus was not punished because he died almost immediately after it was published.

Giordano Bruno, however, was burned at the stake in 1600 for having the same beliefs. Additionally, Galileo Galilei[xvi] published his book on the Heliocentric hypothesis in 1632; he was tried and convicted by the Inquisition, his book added to the index of Prohibited Books, and he was put under house arrest. They were lenient because Galileo stated that he had not intentionally violated the injunction.

With the confluence of religion and scientific consensus, a scientific belief based on bad data, faith, and religious principle can last for a long time; even a long time after it is proven to be totally wrong.

The Clovis People

The Clovis people were considered to be the first humans to settle the western hemisphere, about 13,000 years ago. Initially, this seemed like an excellent hypothesis. Clovis sites were found in many places, easily identified by a specific arrowhead design.

Figure 14: Clovis Arrowhead

Once the idea became widely accepted, the reputations of prominent archeologists set in stone, it became literally impossible to fight the consensus. Those archeologists who believed that they had found earlier sites were denounced as frauds, their careers destroyed; they were no longer published in scientific journals, nor able to get the funding they needed to continue their work.[xvii] I was familiar with this situation during the time that additional data was either ignored or attacked as insufficient. Some scientists even hid their results, afraid of the backlash they would get from the supporters of the Clovis Consensus.

It took many decades for archeology to begin to recognize earlier sites and to rethink how humanity first discovered and then populated the western hemisphere. The new discoveries push the first humans in the Americas to from 15,000 to

20,000 years ago[xviii], with one group of researchers saying they've found potential human sites as old as 120,000 years.[xix]

This research is still ongoing with full knowledge that we are still making new discoveries about how and when humanity populated the planet.

We Are Protected from Comets and Asteroids

One of my favorites was the "scientific consensus" that Earth was protected from comets and asteroids by our atmosphere. In the 1950s the scientific consensus was that all Earth's "meteor craters" were really volcanic, this including the famous Barringer meteor crater in Arizona and all of the craters on the Moon.

Although previous scientists had suggested that the craters were actually the result of the impact of a large meteor or comet striking Earth; without the needed scientific proof, that idea was not considered seriously.

It wasn't until 1950 that a young Princeton PhD graduate student, Eugene Shoemaker, when studying the Barringer Crater, began to seriously suggest that this crater was caused by the impact of a meteor.

This young man had an uphill battle, always true when fighting a "scientific consensus." Shoemaker didn't begin to gain recognition and credibility until he and a USGS mineralogist, Edward Chao, discovered a substance in these craters that could only be explained by the ground being struck at high impact from an outside source. Still, there were plenty of geologists that were not convinced.

Figure 15: Eugene Shoemaker

Fast forward to 1993 when Shoemaker, his wife Carolyn, and David Levy were deeply involved in discovering Earth-Orbit-Crossing Asteroids, in order to give the world an early warning should one be on a collision course with Earth. During this research they discovered a string of comets on a collision path with the planet Jupiter.

This would be the first time that science had ever observed a comet hitting a planet. Even at this late date, there were still plenty of geologists who doubted that comets could do much damage to a planet with an atmosphere, especially an atmosphere as dense as Jupiter's.

There was much discussion about what we would be able to see when the string of comets hit Jupiter. The major consensus was that Jupiter would swallow the comet pieces without a trace visible to the Hubble Telescope, trained on Jupiter, just in case.

As the comets struck the planet, explosions the size of Earth were clearly visible every time one of them hit Jupiter.

From that moment on, there was no doubt, Earth was in potential danger of being struck by a major asteroid or comet; resulting in an ELE[10] that could easily destroy the entire human race.

Figure 16: Comet Shoemaker Levy Hitting Jupiter

My final example of a failed hypothesis in the scientific community was the widespread belief that cholesterol was the major cause of heart disease and must be removed from all diets. For over 50 years there has been medical consensus that high cholesterol caused heart disease.

---

[10] Extinction-Level Event

But in the past few years we've had significant research that seems to contradict that consensus. The research ranges from no relationship[xx] at all to a difference between individuals. Regardless, the consensus has been destroyed as we've discovered more and more about how the body works.[xxi]

What is hard for me to believe is that after literally centuries of intensive study of the human body, we are still a long way from a total understanding of the complexities involved in all of the body's systems and how they support and interact with each other.

Yet, climate scientists, who have been researching the climate for less than one percent of the time scientists have spent researching the human body, either believe they have figured out the much more complex array of systems that drive Earth's climate, or they are trying to convince the world that they've figured it out and we better pay attention or there will be a worldwide catastrophe.

Sorry, doomsayers have *always* been wrong; if not about the outcome, definitely about the cause.

Perfect Storm
"Scientific Consensus" 3

Historically, the claim of consensus has been the first refuge of scoundrels; it is a way to avoid debate by claiming that the matter is already settled. ... Whenever you hear the consensus of scientists agrees on something or other, reach for your wallet, because you're being had.

Michael Crichton

The birth of the Perfect Storm "scientific consensus" around AGW really took hold in 1998 when the El Nino of 1997-1998 created global temperatures than had not been seen for over 100 years.

This event seemed to support the climate models that were being developed to predict the temperatures that Earth might experience due the continued use of fossil fuels. In almost every case the models predicted catastrophic global warming, potentially the end of life on the planet. The timeline and early results varied but the conclusion seemed inevitable.

The scientists involved, committed (as most scientists are) to their models, believed that we didn't have years to do the proper testing and even though they knew that some facets of the hypothesis would probably need to be adjusted, they were convinced that the basic concept was correct. Earth was in trouble because of human activity and if something wasn't done immediately, the next generation would not live out their lives; they would die in a world too hot to maintain human life. For them, this time the sky was really falling!

They screamed their findings and great concern to the public through the media and to the governments of the world through a variety of organizations including the IPCC (International Panel on Climate Change) which has the responsibility of providing "policymakers with regular scientific assessments on the current state of knowledge about climate change."[xxii]

As far as many were concerned, there was no doubt, if we didn't act immediately, all was lost.

The panic that set in has continued unabated to the time of this writing. The true believers, as you will see by the many dire predictions that are still being made, all speak of the terrible consequences if humanity doesn't act immediately.

Here's a list of some of the many predictions where opposing views are not even considered seriously by the elite, the influential AGW fanatics:

- *The sea levels are rising at an alarming rate and could rise as much as 20 feet in a couple of generations.*
- *Major storms, hurricanes, and tornadoes will increase at an alarming rate.*
- *Polar bears are dying because of the temperature increase in the Arctic.*
- *The Arctic is melting. In 2009, Al Gore, the High Priest of Global Warming, stated, with a 75 percent probability that the entire Arctic would be ice free in summers by 2013.*
- *In 2009 NASA scientist James Hanson, of hockey stick fame, stated that we only have four years to save Earth.*
- *In the same year, Canada's Green Party wrote that we only had hours left to stop global warming.*

- *Also in 2009, United Kingdom Prime Minister Gordon Brown said there was only 50 days left to save Earth.*
- *The U.N.'s top climate scientist, Rajendra Pachauri, said in 2007 we only had four years to save the world.*
- *Environmentalists warned in 2002 the world had a decade.*
- *In 1989 Noel Brown, a senior environmental official, stated that entire nations would not survive rising sea levels and that we must reverse this trend by the year 2000 or it would be too late.*
- *The United Nations warned in 2005 that the effects of global warming would lead to massive population disruptions as areas of the world became uninhabitable.*
- *The end of snow forever was predicted.*
- *All of this cold weather is a result of global warming.*
- *Climate scientists predicted that warming temperatures due to CO_2 emissions would cause the polar ice caps to melt.*
- *The belief that ocean acidification[xxiii] is essentially an additional byproduct of the continued use of fossil fuels and that it will cause major problems for people dependent on the ocean for their livelihood.*
- *The coral reefs, especially the great barrier coral reef in Australia, are dying because of the warming seas caused by Climate Change and the burning of fossil fuels.*

Much of what I've reported above came from an excellent article in the New American.[xxiv] The rest comes from years of researching Sky Is Falling frauds and ELEs (extinction level events).

An even closer look at the number of issues laid at the feet of "climate change" and, therefore, the use of fossil fuels for energy, makes one wonder if anything is going wrong that can't be blamed on fossil fuels.

Why did this happen?

Researchers discovered very quickly that if they added to their research proposals that they were going to determine the negative impact of AGW on whatever they were studying, the ability to gain funding was almost automatic. Essentially, almost everyone researching the terrible consequences of AGW found themselves in funding heaven.

The United States alone was spending a billion dollars a year on any research supporting the AGW hypothesis. Plus there were the many private organizations that jumped on the AGW bandwagon and provided additional funding in the tens of millions. Scientists, scientific organizations, and research universities around the world became almost totally dependent on the AGW hypothesis. *It's very difficult to kill the goose that is laying the golden eggs.*

This is essentially the state of affairs even today. The fact that research has been carried out that strongly suggests that most, if not all, of these predictions are basically wrong, has not caused the main forces behind the continuation of the AGW scientific consensus to change their minds. In fact, if anything, it has caused them to dig in even deeper.

So, how did we get here, how did this AGW myth gain such a firm foothold? Here's what happened and why we are in the shape we are in today.

Initially, the idea that fossil fuels were playing a role in the warming of the planet made an excellent hypothesis. There's no doubt that we were putting more and more CO_2 into the atmosphere and it was also clear that the planet was warming. So, the AGW

(Anthropogenic Global Warming)[11] hypothesis was created. This hypothesis stated that as we pumped more and more CO_2 into the atmosphere by burning fossil fuels, the global temperature would rise accordingly.

In 1999 Mann and associates took this concept to an entirely new level with models and the hockey stick graph showing that the warming would not only increase, but increase exponentially. This is a scary proposition and is best explained with this scenario.

Let's say a pond has one cell of algae in it and that algae and all future cells of algae divide once a day. If the pond is completely full of algae on day 30, when is it half full? Obviously, on day 29.[12]

This potential screams about the danger of global warming if we wait too long. If we wait until day 29, we will have waited too long. Because of this, the urgency was shouted to the world before the hypothesis could be validated.

This action completely nullified the scientific process that demands proof of concept before any hypothesis is proven to be correct. This proof of concept involves thorough testing by other researchers who must achieve the same results and, most importantly, any models created to support the thesis must accurately predict future results.

---

[11] Remember, for the purposes of this book we are taking the AGW fanatics definition of AGW. This is basically that human activity essentially through the use of fossil fuels is bringing about catastrophic global warming that will threaten, if not end, life on this planet.

[12] Paul Hawken (http://www.paulhawken.com/) was the first person to present this scenario to me, that was 25 years ago when we were working together on the *Seize the Day* calendar project.
http://www.iangilman.com/software/seizetheday.php)

It is interesting that it is the "warmists,"[13] who are screaming that realists are anti-science, when it is their behavior that is totally ignoring the scientific process.[14] To be fair, they have their reasons; most believe their actions are essential to save the planet and those living on it. Some are just plain charlatans, supporting AGW to further a private agenda.

The idea the world was about to end and that humanity was the cause appealed to many groups: environmentalists, the mainstream media, progressives/globalists, governments, entrepreneurs, and ultimately a large portion of the scientific community.

Environmentalists thought they'd died and gone to heaven. Finally the world was recognizing the danger humanity was to the planet. They'd always believed this and some of the real nuts in this community had actually become eco-terrorists. According to the FBI, eco-terrorists caused hundreds of millions of dollars of damage between 2003 and 2008. Regardless, the scientific hypothesis that humanity was destroying the planet was music to the ears of environmentalists across the globe.

The mainstream media was almost giddy. This was the story of the millennia, only the Second Coming could outdo it. The world was going to end, the sky was actually falling, and, best of all, we were at fault through our greed, our raping of the environment, our ignoring the signs that we were going too far, testing the ability of Earth to handle our misdeeds. They couldn't get enough of this bad news, nothing sold media any better. The last thing they wanted to hear was that the hypothesis was wrong. Not only would they have a dozen eggs on their respective faces, but the story

---

[13] Term used to identify fanatics who see the AGW hypothesis literally as gospel
[14] This is a standard technique of the progressive movement and individual progressives (the loudest and most dangerous supporters of AGW).

would die and they would never have another one so good, so juicy, so deliciously bad.

Progressives and globalists embraced the hypothesis with open arms. This was a problem only governments could solve, in fact, a worldwide government. They had been dreaming of this potential for decades, if not centuries. The ultimate power of a worldwide central government that could lay down the law, make decisions that the *hoi polloi* could not be trusted to make. Ultimate power more than the pharaohs of Egypt could even dream of.

The progressives didn't want a constitution that limited the power of the central government. They wanted to be able to expand the power of the central government to the extreme, and the AGW hypothesis gave them the excuse they needed to demand that the central government take over every facet of life on Earth.

Entrepreneurs saw an entirely new industry being created, the green industry. Consultants would be needed to judge whether a building met the new "green" requirements. Solar power, wind farms, electric cars, battery technology, etc., etc. would be needed. And the need was filled, the AGW hypothesis created an industry that in 2018 was worth 4 trillion dollars. It is estimated that this industry will be worth 90 trillion dollars by 2030.[xxv]

With the AGW hypothesis governments finally had "reasonable" reasons to increase taxes (hidden taxes) on the poor and middle classes. They might make a showing by lowering income taxes or providing certain tax breaks for those classes. But they could soak them by taxing energy companies (the carbon tax), who would be forced to pass the added expenses on to their customers. Thus, the poor and middle classes would suffer the most because of the resulting increases in the price of energy for cooking, heating, cooling, driving, and on and on.

This never ends because everything we consume depends on energy either to produce it or deliver it.

Tax increases are only one of the boons the federal government gains. They also gain immense power over people's lives as is best demonstrated by the massive rules and regulations created by the Obama EPA. Power over the people, finally the end of that ridiculous old-fashioned saying of a "government of the people, by the people, for the people, shall not perish from this earth."

Finally, those with the knowledge and skills to run the world, to run other people's lives, would have the opportunity. As Obama has so clearly stated it, "It would be so much easier to be President of China."

With these groups firmly on the side of the AGW hypothesis, things began to develop rather rapidly. The people wanted to know what global warming meant in their lives and they turned to the government to tell them.

Governments gladly took on the responsibility and began handing out what ultimately was billions of dollars to research facilities that were willing to embrace the hypothesis and answer the question, what does this mean?

As I stated earlier, we've had scientific consensus before and in almost every case they tend to hamper research, setting it back decades if not centuries, destroy reputations, cut off the funding of those opposing the consensus, and, ultimately, they are proven to be wrong.

But, this was different, this was the first time in literally centuries that the consensus had the support of governments, the media, and the people. The last time this happened was when the consensus insisted that Earth was the center of the Universe; that time people were burned at the stake, not allowed to publish or lecture, and scientific research was put on hold for decades (actually centuries).

    This time it's even worse, this time we are regressing to a time when "science" was what the "priests" believed and no opposing voices were allowed to be heard.

This time we have the Perfect Storm "scientific consensus."

The Opposing View   4

All scientists should be skeptics. The reason why is that, even with the best of scientific measurements, we can come up with all kinds of explanations of what those measurements mean in terms of cause and effect, and yet most of those explanations are wrong. It's really easy to be wrong in science ... it's really hard to be right.

Roy Spencer

So that's how it all began, but there is another side of the story, the opposing side, those of us private citizens, scientists, and politicians who are denigrated and called deniers. Why do we doubt the consensus? That's what this chapter is all about.

The models suggesting AGW were based on an untested hypothesis; the proper scientific process was completely ignored and a dangerous (to human life and our nation's socioeconomic health) hypothesis was allowed to influence governmental policy around the world, from the smallest countries to the U.N. itself.

The way science happens is that scientists notice something and that sparks their curiosity.

I remember reading Dr. Richard Feynman's book, *Surely You're Joking, Mr. Feynman!* In it he describes his inspiration for one of his discoveries. He was at a circus-theme restaurant having dinner and as he watched the professional plate-twirler, twirl plates on the end of a stick, he noticed a wave pattern that he'd never seen before. His curiosity was peaked and the research that he did to satisfy his curiosity led to his receiving the Nobel Prize for Physics.

Again, the scientist notices something that sparks their curiosity. After some research a hypothesis is formed that attempts to explain what has been observed. Then models are created that are designed to test the hypothesis.

Once the hypothesis is proven to the scientist's satisfaction, a paper is published; usually in a peer-reviewed scientific journal that defends the hypothesis, fully describes the research, and provides all of the necessary data.

This is done so that other scientists can have a crack at it, try to duplicate the results. If it passes muster, the hypothesis gains some credibility, but it is still not considered to have reached the level of an acceptable theory, definitely not a scientific fact.

I have often been told that all important research is peer reviewed. There are two problems with this concept. First, when you have a scientific consensus, it is not uncommon for the opposing view to be shut out of the review process. I know this is hard to believe, but it is a fact of scientific life. Additionally, even if the opposing research is published, other scientists don't want to get involved; they don't want their name attached to a project that is counter to the consensus.

Second, there are massive problems[xxvi] within the peer-reviewed industry. Those involved at every level are saying that the process is too slow and too expensive, that it produces inconsistent results, that often personal agendas and bias create results that cannot be trusted.

Finally, the process itself can be and is abused in many different ways. For example, a reviewer can give a competitor a very harsh, unreasonable review or purposely slow the process down, so a competitor will miss an important drop-dead date for funding.

There are times when the honest scientist discovers the hypothesis is completely wrong, even if it has been an accepted hypothesis. For example, the 2011 Nobel Prize for Physics[xxvii] was awarded to Saul Perlmutter, Brian P. Schmidt, and Adam G. Riess for discovering that the rate of expansion of the Universe was increasing.

The interesting thing was that their research was designed to determine how much the expansion rate was slowing. This made perfect sense. The Universe's expansion began about 13 billion years ago; the odds that it was finally slowing down were pretty good. However, they actually proved the opposite and the results surprised them as much as it surprised the scientific world. The proof didn't come easy; they had to run their tests and calculations over and over again. In addition, they needed verification from other researchers. The results kept coming out the same; the rate of expansion of the Universe was increasing.

This is an excellent case of scientific research being done honestly, of not being afraid to challenge a current belief, no matter how much acceptance that belief has. When the data seems to be showing that something else is going on, another unbiased look is needed if the truth is to be discovered.

In fact, recent researchers, using a larger data set, have determined that the rate is actually constant.[xxviii]

In science, everything always changes, nothing stays the same.

This essential, scientific process was ignored in the case of the AGW hypothesis. Again, the climate "scientists" had, what was for them, a reasonable, even responsible, excuse. There wasn't time to wait for the hypothesis to be proven scientifically. That would take at least a decade and could take even longer. They "knew" that we didn't have a decade; the hockey stick graph showed that global warming was on an exponential curve. There was literally no time left; the train was leaving the station, the conductor had sounded the alarm, and the world jumped on board for a trip to fantasyland.

What follows is current, opposing research of some of the significant AGW issues.

As you read what I've reported, ask yourself this question, *"Shouldn't I give as much credence to scientists whose only efforts are to discover the truth, as I give to scientific organizations, journals, research universities, and individual scientists whose reputations and funding are almost entirely due to their support of the AGW hypothesis?"*

Opposing Research Regarding Significant AGW Issues

Sea-Level Rise

"The sea levels are rising at an alarming rate and could rise as much as 20 feet in a couple of generations."

– James Hansen

There's no doubt that sea levels have been rising; rising for the past one hundred plus years, the result of the mini ice age of the late 1700s and early 1800s; however, regardless of what you read in the progressive press, the rate of rise has not increased.

Dr. Nils-Axel Morner of Stockholm University, formerly chairman of the INQUA International Commission on Sea-Level Change, has attacked the IPCC report predicting major sea-level rise. His research shows a maximum rise in sea levels this century of 4 inches. He is appalled that the IPCC would suggest anything greater and is convinced that the reports of massive sea-level rise are nothing more than "scare" stories, akin to a child's fear of monsters under their bed.

Morner is not alone by any means.

It appears that Earth's coasts actually gained land over the past 30 years, according to another study published in *Nature Climate Change*.

Researchers led by Gennadii Donchyts from the Deltares Research Institute in the Netherlands found that Earth's surface gained a total of 58,000 square kilometers (22,393 square miles) of land over the past 30 years, including 33,700 sq. km. (13,000 sq. mi.) in coastal areas.[xxix]

"We expected that the coast would start to retreat due to sea-level rise, but the most surprising thing is that the coasts are growing all over the world," study co-author Fedor Baart told the BBC.

Finally, the world's media is keeping alive the myth that the Marshall Islands are disappearing because of sea-level rise due to AGW.[15] However, the research done by Murray R.Forda and Paul S. Kench and reported in *Science Direct*, shows that in the Marshall Islands, there has actually been land gained from the sea, not lost to the sea, over the last 60 plus years.[xxx]

However, this myth is alive and well today with the "scientific consensus" in full support of the catastrophic predictions being made. The fact that this just isn't happening doesn't stop the consensus from continuing to publish outlandish predictions.

---

[15] For example, a Google search of "Marshall Islands disappearing" delivers 697,000 results.

Major Storms, Hurricanes and Tornadoes Will Increase at an Alarming Rate.

"All across the world, in every kind of environment and region known to man, increasingly dangerous weather patterns and devastating storms are abruptly putting an end to the long-running debate over whether or not Climate Change is real."[16]

–Barack Obama

Again the consensus completely ignores reality; the incidence of tornados has been steadily decreasing for the past 45 years. Not only did Mother Nature just set a record for lack of tornado activity, she absolutely shattered the previous record for fewest tornadoes in a 12-month period. During in the 2012 – 2013 tornado season only 197 tornadoes struck the United States. Prior to this past year (2018), the fewest tornadoes striking the United States during a 12-month period occurred from June 1991 through July 1992, when 247 tornadoes occurred.

Since 2005 we experienced the longest hurricane drought in history. It finally ended in 2017, 12 years without a hurricane making landfall in the United States.[17]

In addition, some are very concerned about what appears to be an increase in the number of, and damaged caused by, forest fires. They blame this "increase" on global warming. However, a University of Washington climate scientist, Cliff Mass, did the necessary research and discovered that there had been no increase in wildfires in California for decades.[xxxi]

---

[16] Of course Climate Change is real. I'm sorry, this is an ignorant statement and I don't give a damn who said it.

[17] When Sandy made landfall, even though it caused tremendous damage, it was only a tropical storm.

In addition, according to the National Interagency Fire Center, there were 67,743 wildfires in 2016. That's down from more than 85,000 in 1986. By December 22 of last year (2018), there had been about 66,000 fires, NIFC data shows.[xxxii]

Polar Bears Are Dying Because of the Temperature Increase in the Arctic.

"For humans, the Arctic is a harshly inhospitable place, but the conditions there are precisely what polar bears require to survive - and thrive. 'Harsh' to us is 'home' for them. Take away the ice and snow, increase the temperature by even a little, and the realm that makes their lives possible literally melts away."

—Sylvia Earle

This is either a clever lie designed by an immoral PR individual to get the children on their side, so children can help convince their parents as to the "fact" of AGW; or, it's poor research on the part of the "scientists" responsible for the panic.

The true population of polar bears is difficult to ascertain; however, today the estimate worldwide is approximately 30,000. That's a 30 percent increase since 2005.[xxxiii]

Do not believe that polar bears are under any threat due to any climate crisis. In fact, research shows that polar bears have already lived through three climate changes where the Arctic was much warmer than it is expected to be during this current period of planetary warming.

The Arctic Is Melting

The entire North Polar Ice Caps during some of the summer months could be completely ice free within the next 5 to 7 years.[xxxiv]

—Al Gore

In 2009, Al Gore, the High Priest of Global Warming, stated, with a high probability, that the entire Arctic would be ice free in some summers by 2013. This is only one of his spectacular misses; he's really got nothing right. Yet he is still highly revered within the environmental community and the media. Will they ever learn?

Actually, during October 2013, sea-ice levels grew at their fastest pace since records have been kept, and September 2016 was another record-breaking month.

The real record breaker was the March ice measurements of 2019, compared to the past 15 years. As you can see in the chart below, the sea-ice volume increase in March of 2019 blew away any previous sea ice measurements.

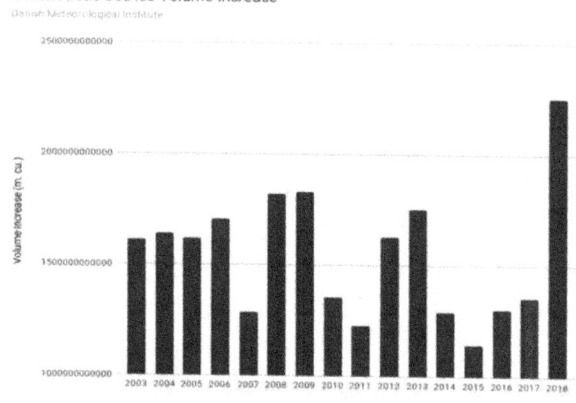

Figure 17: Arctic Sea Ice Measurements

It seems as if Mother Nature is playing with these "experts." Every time they make a prediction or have a meeting, she makes them look like the fools they are. When are they going to realize that Climate Change is a much more complex issue than their models will ever be able to represent?

Only a Few Years to Save Earth

In 2009 NASA scientist James Hanson, of hockey stick fame, stated that we only have four years to save Earth.

Hanson has as great an ego as Gore and appears to be just as wrong. What these faux scientists do when their statements are proven false is to state they just had the timing wrong, but that their basic thesis is spot on. Recently they have gotten wiser; their drop-dead now is about 50 years out. In this way they can escape the ridicule of being wrong while they are still alive.

But he was far from the only one to scream that devastating prediction.

- *In the same year, Canada's Green Party wrote that we only had hours left to stop global warming.*
- *Also in 2009, United Kingdom Prime Minister Gordon Brown said there was only 50 days left to save Earth.*
- *The U.N.'s top climate scientist said in 2007 we only had four years to save the world.*
- *Environmentalists warned in 2002 the world had a decade.*

Areas of the World Will Become Uninhabitable

The United Nations warned in 2005 that the effects of global warming would lead to massive population disruptions as areas of the world became uninhabitable.[xxxv]

According to the report we would see these results within 10 years. What we've actually seen is that humanity is very good at "massive population disruptions," not from global warming but from wars, terrorists, and evil rulers.

There's a story in *The Atlanta Journal, Constitution* (2/17/17), where Gore was giving the exact same prediction to a group of global warming fanatics in Atlanta, Georgia. He is the Chief Priest of AGW; luckily, he hasn't sacrificed any enemy warriors, but I'm sure that he'd love to.

Actually, many areas of the world are becoming more habitable as the increase in atmospheric CO_2 greens the planet.[xxxvi]

The End of Snow Forever

The end of snow forever. Climate alarmist David Viner said in 2000 that snow would soon be a thing of the past, that our children would have to read about snow in history books. The past winters in both the United States and Britain have proved him to be another Climate Change (global warming) charlatan. Although, he probably believes that he just had the timing wrong.

Cold Weather is a Result of Global Warming

All of this cold weather is a result of global warming. This idiocy has been touted by no one less than the White House Science "Czar" John Holdren, under the Obama administration. This is standard practice amongst progressive scientists or politicians. When the facts don't support your initial prediction, blame the results on someone or something else. However, blaming unexpected extremely cold winters on global warming is taking this idiocy too far.

Ocean Acidification

"Evidence gathered by scientists around the world over the last few years suggests that ocean acidification could represent an equal — or perhaps even greater threat — to the biology of our planet than global warming."

- Professor Ove Hoegh-Guldberg

This is another issue that is treated as an absolute fact by all of those who support AGW. In fact, it is often used as proof of the actual danger of human-caused catastrophic global warming.

But, as usual, there are opposing views by scientists who have studied this issue for years.

Howard Browman in his *ICES Journal of Marine Science*, "Applying organized skepticism to ocean acidification research,"[xxxvii] is extremely concerned that the proper scientific journals are not allowing opposing views to be published and, therefore, the scientific community and the public are only aware of one side of the issue.

(Research articles) typically published in "high impact" journals and covered by the mass media, predict an OA-generated ocean calamity. ... *As is true across all of science*[18], studies that report no effect of OA are typically more difficult to publish and, when published, seem to appear in lower-ranking journals. ... For these reasons, the *ICES Journal of Marine Science (IJMS)* solicited contributions to a special issue with the theme, "Towards a broader perspective on ocean acidification research".

Although I call for a more sceptical scrutiny and balanced interpretation of the body of research on OA, it must be emphasized that OA *is* happening and it *will* have effects on some marine organisms and ecosystem processes.

---

[18] Italics mine

Browman sites many research papers that are in opposition to the hypothesis that has identified the problem as catastrophic. However, as he clearly states, it is very difficult to get published when you are going against the "scientific consensus"; this means that articles in opposition to the consensus often appear in the general press. Here are some examples:

Probably the best explanation of why Ocean Acidification is a myth is found in the *Watts Up With That* (WUWT) article, "The Total Myth of Ocean Acidification,"[xxxviii] by David Middleton.

In this excellently researched article he shows clearly that Ocean Acidification is essentially impossible and that any suggestion of that end point is essentially a lie designed to manipulate the opinions of the ignorant, concerned people.

> Just as we describe an increase in temperature from -40°F to -20°F as warming, even though neither the starting nor the ending temperature is "warm" the term "acidification" describes a direction of change (i.e. increase) in the level of acidity in the global oceans, not an absolute end point. When CO_2 is added to seawater, it reacts with the water to from carbonic acid (H_2CO_3; hence acid is being added to seawater, thereby acidifying it.

However, the seawater is still alkaline in nature.

He goes on to say that current scientific research is pretty clear, ocean acidification is a non-problem and no matter what happens, things are pretty much going to stay the same.

Jonathan DuHamel in the article, "The Myth of Ocean Acidification by Carbon Dioxide"[xxxix], published in *Arizona Independent News Network*, presents a strong case to support his contention that ocean acidification is essentially nonexistent, a non-problem.

It has been estimated that current ocean pH is 0.1 pH unit less alkaline than it was in recent pre-industrial time, and some climate models predict a further decrease of 0.7 pH units by 2300.(2) However, proxy reconstructions of ocean acidity, based on fossil and modern corals, show that ocean pH has oscillated between pH of 7.91 and 8.29 during the past seven thousand years.(3) That cyclic variation is nearly four times larger than the 0.1 decrease alarmists are whining about, and even if the model predicted decrease of 0.7 units occurs, the water will still be alkaline.

There seems little doubt that the alarmists have exaggerated the ocean acidification issue for their own purposes, have essentially lied to the general public in order to manipulate their beliefs regarding Climate Change and its impact on Earth's ecological system. This is not a surprise; they have been exaggerating AGW issues for over three decades. Why would they stop now?

Returning to Howard Browman's concern, the entire scientific community, especially those scientists involved in climate research, must acknowledge the need for more skeptical scrutiny and balanced interpretation of the body of research. This need is absolutely essential in regard to AGW and, as you will see in the addendum, also in regard to human-caused massive extinction.

Coral Reefs

This is what NOAA has to say about coral reefs and Climate Change.

> Climate Change is the greatest global threat to coral reef ecosystems. Scientific evidence now clearly indicates that Earth's atmosphere and ocean are warming, and that these changes are primarily due to greenhouse gases derived from human activities.

As temperatures rise, mass coral bleaching events and infectious disease outbreaks are becoming more frequent. Additionally, carbon dioxide absorbed into the ocean from the atmosphere has already begun to reduce calcification rates in reef-building and reef-associated organisms by altering seawater chemistry through decreases in pH. This process is called ocean acidification.

Climate Change will affect coral reef ecosystems, through sea level rise, changes to the frequency and intensity of tropical storms, and altered ocean circulation patterns. When combined, all of these impacts dramatically alter ecosystem function, as well as the goods and services coral reef ecosystems provide to people around the globe.

Again, is the danger as great as NOAA is stating as absolute fact?

Jennifer Marohasy[19] in the article, "Reefs may benefit from global warming,"[xl] published in *Science Alert*, is clear in the accuracy of her research into the global warming/coral reef issue.

> The idea that the Great Barrier Reef may be destroyed by global warming is not new, but it is a myth. The expected rise in sea level associated with global warming may benefit coral reefs and the Great Barrier Reef is likely to extend its range further south.

> Most of the world's great reefs are tropical because corals like warm water. Many of the species found on the Great Barrier Reef can also be found in regions with much warmer water, for example around Papua New Guinea. Corals predate dinosaurs and over the past couple of hundred million years have shown

---

[19] She was a senior fellow at the free-market think tank the Institute of Public Affairs between 2004 and 2009 and director of the Australian Environment Foundation until 2008.[li] She holds a PhD in biology from the University of Queensland.

themselves to be remarkably resistant to climate change, surviving both hotter and colder periods.

Interestingly, scientific studies show that over the past 100 years, a period of modest global warming, there has been a statistically significant increase in growth rates of coral species on the Great Barrier Reef. There have also been periods of coral bleaching, but no conclusive evidence to suggest that either the frequency or severity has increased.

Once again, it seems obvious that "Having eyes, do you not see? And having ears, do you not hear?" The science is not settled, the "scientific consensus" notwithstanding.

To quote Einstein, "Why 100? If I were wrong, one would have been enough." Or, to paraphrase, "A 'scientific consensus'? One person sitting at the kitchen table could prove me wrong."

Ultimately, of course the AGW hypothesis was tested and honest, accurate, scientific data proved that global warming essentially halted in 1999. Even though 1998 was one of the warmest years on record (due to a very powerful El Nino), temperatures have not increased at anywhere near the rate that the hypothesis predicted, definitely not exponentially, and effectively, really not at all.

In addition, the Antarctic ice sheet is growing and the Arctic has yet to see a summer without ice. There's still plenty of snow around, with last year showing records for California, the upper Midwest, and the East Coast.

Powerful storms seem to be declining, again, definitely not increasing. And, let's not forget the polar bears who have enjoyed a great resurgence since the late 1960s when the polar bear population was about 12,000; the latest estimates put the population at about 39,000 worldwide. In addition, as shown earlier, it appears that polar bears have survived, over the past million-plus years, periods of warmth much greater than we will experience in the next decades and probably much longer.

Remember, trillions of dollars are at stake, as well as the reputations and credibility of many scientists, scientific organizations, and other groups of individuals. There is nothing that will make a person ignore the truth faster than a fact that will take dollars out of his pocket. A fact that will take all his dollars and his reputation, too, is a fact that is extremely difficult not to ignore.

As realists, our only issue is that we can't ignore the scientific process in order to support an unproven hypothesis. When government policy is created, based on an unproven hypothesis, that adds insult to injury. And, when those with an opposing view are shut out of the conversation, that's adding potential disaster to injury, and nothing good ever results.

Unintended Consequences   5

From Torquemada to Robespierre and Hitler the men who have made mankind suffer the most have been inspired to do so by a strong faith; so strong that it led them to think their crimes were acts of virtue necessary to help them achieve their aim, which was to build some sort of an ideal kingdom on earth.   David Cecil

This is not just a scientific argument. It has become a major societal decision that will affect millions of people's lives in the United States, and ultimately the lives and well-being of billions living on Earth. There are at *least* six potential dangers to staying on the AGW hypothesis path:

1) The loss of critical research carried on by scientists who doubt the predictions made to support the AGW hypothesis
2) Damage to our nation's socioeconomic health
3) Suffering and loss of life worldwide
4) Being unprepared should the world enter a mini ice age
5) Lessening the amount of CO_2 in the atmosphere resulting in declining plant life and production of atmospheric Oxygen
6) Delaying the discovery of the actual processes that drive climate change
7) And, the rise of a world government; the end of nation states

The Loss of Critical Research Carried on by Scientists Who Doubt the Predictions Made to Support the AGW Hypothesis.

Those in the scientific community who were concerned that the hypothesis was still unproven were shut down and didn't receive any government funding. What funding they did get from organizations that wanted to know what was really going on, quickly dried up as public opinion turned against those that opposed the hypothesis.

Any researcher that desired to test the hypothesis and/or offer opposing views was taken to the woodshed or made to sit in the very back of the bus. In essence, it was a career-destroying decision to go against what had become not only a scientific consensus, but a "perfect storm" scientific consensus. This powerful focus on AGW has pushed climate science back decades and the idiocy seems nowhere near ending.

In addition, the billions spent trying to demonstrate the disaster AGW is visiting upon the environment and, therefore, humanity, are dollars that could have been spent on other much more pressing issues.

Bjørn Lomborg[xli] probably understands better the worldwide economics of the overemphasis on AGW than anyone else. Even though he sees global warming as a potential threat (however, not at the AGW level) he believes that the financial resources and time spent on AGW would have been much better spent in other areas. Lomborg says:

> The World Bank does a lot of important and effective work, especially in health and education, but its climate policies are poorly considered. The Bank's new president, David Malpass, should refocus the

institution on its core mission of eradicating poverty – including the energy poverty that wrecks too many lives.

Damage to Our Nation's Socioeconomic Health

The many billions of dollars that must be spent to, theoretically, halt global warming will greatly damage our socioeconomic health through the massive taxation that will be necessary to support the effort. This effort will have to be borne by the poor and middle classes because the taxation will be a tax on energy (carbon tax) and the energy companies will pass it on in the form of price increases. The government knows this but assumes that the tax payer will blame the energy companies; this is why it is call a "hidden tax."

It is a fact that the life expectancy in the United States went down a bit in 2018. The question, of course, is why did that happen? The answer is obvious to those of us who understand the power of images, the power of the negative images being spread by men like Ehrlich, the media, progressive politicians, Hollywood, etc. The despair, intense fear of the future created by these influential sources is causing an increase is drug deaths and suicides; this is bringing down life expectancy in the United States. In essence, the doomsayers are bringing about that which they most fear; they are creating a self-fulfilling prophecy, pointing at it and stating, "See … We told you this would happen."

Suffering and Loss of Life Worldwide

What major development has taken a major segment of humanity from day-to-day suffering to a relatively comfortable successful life; has allowed us to feed more people than experts ever believed possible? The answer is simple: the energy that comes from fossil fuels. Fossil fuels drive the engines of progress and make the world a safer place to live. They bring light to the darkness, give warmth in the winter and cool on a hot summer day. They bring food to the table and medicines to the hospitals. They provide a level of individual freedom to billions of people that, not too long ago, was only available to the very wealthy.

Lomborg[20] again,

> Fossil fuels do contribute to global warming – but also to prosperity and well-being. One billion people worldwide live in homes that lack the energy to light even a single bulb. And more than three billion live in countries without reliable, 24-hour energy networks that can power hospitals and factories.

> At the same time, three billion people currently suffer from terrible indoor air pollution, because poverty forces them to burn dirty fuels such as wood and dung to cook and keep warm. But solar panels cannot power clean stoves or heaters, or refrigerators that would stop vaccines and food from spoiling. Nor can they power the agricultural and industrial machinery that supports jobs and pathways out of poverty. In that respect, distributing solar panels is mostly a way for rich people to feel good about taking action on global warming.

---

[20] Lomborg has studied this issue for over a decade. In 2007, he was named one of the 100 Most Influential People by *Time* magazine after the publication of his book *The Skeptical Environmentalist,* which challenged widely held beliefs that the environment is getting worse; that AGW existed.

While supporting the basic concept that human activity plays a role in global warming, Lomborg realized that there was basically nothing that we could do about that, and that the trillions of dollars needed would be much better spent in other areas which include research in technologies that will minimize the impact without disrupting life on the planet. I encourage you to do a Google search and learn about some of the ideas that this brilliant man has to "save the planet."

The hope was to spread this boon of affordable energy throughout the world so that everyone would at least have light, warmth, and a hot meal and possibly everything else that energy can bring.

    The need is great, millions are dying from the burning of biofuels to cook and heat homes. These deaths are caused when the major source of energy for heating and cooking is biofuels (wood, cow dung, etc.). Burning bio fuels directly, in the home, is causing hundreds of thousands of deaths around the world every year. This could easily be solved by providing electricity for heating and cooking to every person on the globe.

    Currently this can only be accomplished through the use of fossil fuels. However, since there is continued pressure not to provide the energy sources and distribution technologies needed; those dependent on biofuels will continue to suffer approximately 200,000[xlii] deaths a year.

    But, warmists don't care; all they care about is their precious belief that humanity is destroying the planet. If this attempt to rid the world of fossil fuels continues, there will be many others, even in countries like the United States, that will suffer a loss of freedom and, potentially, loss of life if they can no longer afford to buy the energy they need to heat their homes and cook their food.

Bottom line, if we toss AGW under the bus (where it belongs), we know how to provide the energy needed without causing air pollution. We've done that in the United States, and we can do it around the world.

Being Unprepared Should the World Enter a Mini Ice Age

Solar scientists[xliii] around the world are verifying a link between a "Quiet Sun" (when the Sun has a lack of sunspots) and global cooling. This link has been known for decades because we have sunspot data that goes back centuries and climate data to associate with that.

Our Sun is currently entering one of its greatest solar minimum phases that science has ever observed and some solar scientists are predicting that a mini ice age will begin in 10 to 20 years. If they are correct, this will result in shorter growing seasons and a much greater need for energy to heat homes and businesses and cook whatever food is available.

China seems to recognize this potential and they are buying up fossil fuel energy sources around the world, plus adding many more coal plants and refineries. This fact is not well known because the warmists don't want it known that China's commitment to the AGW hypothesis is not anywhere near as strong as we are being led to believe.

In other words, they are making the right decisions and we are in danger of making exactly the wrong ones. Our efforts to halt Climate Change are exactly the opposite of what we would be doing if we recognized the potential of a mini ice age within a couple of decades. Currently we are not prepared for this eventuality and the suffering that could result will not be pleasant to experience nor observe.

If this proves to be true, the impact on the worldwide availability of food would be devastating. I'm not trying to create a Sky Is Falling scenario, but this is something we could prepare for.

However, currently, our efforts to stop global warming are in direct opposition to what we would need to do if we end up in another mini ice age.

Essentially we would need to put some grain crops in storage, and quit using corn to create ethanol. In addition, we would need to garner all of the energy sources we could so that we could ride out the cooling in relative comfort.

As you can easily see, shutting down coal mines and oil/gas fields while shifting too prematurely to alternative fuels is exactly what we would *not want to do* under the cooling scenario. In my opinion, the cooling scenario is significantly more probable than AGW.

Lessening the Amount of CO_2 in the Atmosphere

More CO_2 produces hardier plants that resist disease and the effects of pollution. With the increase of CO_2 there has been a significant greening of Earth[xliv], with forests and jungles thriving. This is where the environmentalists lose me. They have been screaming for years that our forests and jungles are in great danger, but when we discover that the increase in CO_2 is almost solving that critical issue, this information is ignored and they are unable to grasp anything positive about the situation.

NASA has also noticed that Earth is greening significantly:

> An international team of 32 authors from 24 institutions in eight countries led the effort, which involved using satellite data from NASA's ... instruments to help determine the leaf area index, or amount of leaf cover, over the planet's vegetated regions. The

greening represents an increase in leaves on plants and trees equivalent in area to two times the continental United States.

There are many benefits[xlv] of adding CO_2 to the atmosphere but this fact is given little press from the progressive media. For example: The productivity of food crops is significantly enhanced as are many of the beneficial substances, i.e. vitamins and antioxidants. Our food promotes better health naturally. The photosynthesis of all plants: farm crops, forests, jungles, even weeds is increased ultimately resulting in a greater abundance of plants and, therefore, more oxygen. CO_2 is essential to plants; oxygen is essential to animals, including Homo sapiens.

Delaying the Discovery of the Actual Processes that Drive Climate Change

This is a major issue for the advancement of scientific knowledge and the issue is common every time there is a scientific consensus. The delay in new knowledge almost always lasts for a few decades but there have been times when the delay has lasted for almost a century.

Knowing the actual causes of Climate Change would be extremely valuable and could make the difference in the lives of almost everyone on the planet. We could prepare for periods of extreme cold and take full advantage of periods of warmth. In addition, we might get a sufficient warning of another ice age that would allow us to make the necessary arrangements to minimize the loss of life and human suffering.

This is currently impossible, since almost every penny of research funds is being spent to support the AGW hypothesis and to try and discover ways that we can stop the inevitability of global warming or cooling. Some funds are also looking at technology that can help us adjust to the changes, but this is a minor effort. No funds are being allocated to discover the actual processes and this is literally a crying shame. The Perfect Storm "scientific consensus" will not allow even the thought that CO_2 is not the major, if not the only, contributor to global warming; opposing views are being shut out of the conversation. Well, at least we aren't burning "deniers" at the stake and nobody has gone to jail or been sent to a reeducation camp ... yet!

The Rise of a World Government and the End of Nation States

I've talked a lot about the dangers of a worldwide government. There is little doubt in my mind that it would quickly devolve into a worldwide dictatorship. That is essentially the history of the world. The United States is exceptional because we are the nation that said that government serves the people, that the people do not serve the government.

In every other system of government the people are beholden to the rulers, whether they be kings or queens, dictators or supreme leaders, priests or Imams. It is understood that the people, the masses, serve these rulers and are beholden to them.

The United States is different. We have a nation "of the people, for the people, and by the people" and this concept is at the heart of the Constitution and Bill of Rights. In the United States the federal government's powers are limited and powers not given to the federal government are given to the states.

Our founders understood that a powerful federal government was a parasite feeding on its citizens. If this is true of one nation, imagine the disaster that would result from a worldwide government. Those that support it are either ignorant of history or power-hungry political animals (progressives).

Additionally, if this took place it would guarantee that China would soon rule the world. There's no doubt in my mind that they are planning for the eventuality; however, it may come about. I suspect that in this case there would ultimately be a major conflict between the Muslim world and the communists; quite possibly a nuclear war. The communists would win and the world would suffer centuries of communist rule.

If that happens, the dream of America and the peace and prosperity that we have literally brought to the world will be forgotten. This is not a future that I wish on my progeny.

Finally, the AGW hypothesis has become a religion where God has been replaced by Mother Earth. This religion has all the trappings of any religion, with doctrine that is sacrosanct and where any opposition is considered sacrilege. This religion has its high priests, acolytes, and prophets. Like many religions, followers know the truth and see the unbelievers as ignorant and not worthy until they accept the truth and are converted to the true faith.

Because the true believers are committed body and soul to the AGW hypothesis, the chance of their ever seeing the truth is beyond their reasoning ability. It would be like trying to convert a Jehovah's Witness to atheism, or to convince a creationist that the Universe is billions of years old.

Our only choice is to relegate them to the same level of societal acceptance that all religious fanatics enjoy. They will be free to worship their religion and continue to proselytize, to convince others of the truth of their faith. But, they will not be able to make governmental policy decisions, force America to accept their religion as the religion of the state. We can't afford to have fanatics of any stripe in charge of our nation.

This is much more than a scientific disagreement; this is a war. Who wins will determine whether we are a nation for, by, and of the people; or a nation where the people are beholden to their rulers. I know my preference, what's yours?

Conclusion

To conclude I want to point out an instance where a scientific consensus was turned on its head. Due diligence of the scientists involved in this resulted in a net increase in our knowledge about the Universe and not in the "blacklisting" of the scientists who made the discovery, thereby destroying the scientific consensus that the Universe's expansion wss slowing down.

The consensus existed because it made good sense. The Universe is about 12 billion years old; that's how long ago the Big Bang occurred and the Universe came into existence. That's a long time, by now the expansion must be slowing down; it didn't make sense that the Universe would still be expanding at a faster rate.

However, three astronomers' research provided compelling data that the expansion rate was not slowing.[xlvi] When they analyzed a specific type of Supernova, using the Hubble Telescope, the data proved the Universe was expanding in all directions at once and at an increasing expansion rate. The results were so spectacular and proofs so exceptional that the hypothesis of a Universe expanding at an increasing rate was quickly accepted by the scientific community.

But a few years later, using a much bigger data set, another group of researchers determined that the expansion rate seems to be constant. So, we have gone from slowing down, to speeding up, to remaining constant; can't they make up their minds?

This is very confusing to many people; however, this is exactly how science is supposed to work. This is the process that climate scientists have not only ignored but tried to eliminate. To them what seems right is right, and anyone who disagrees is a denier who must be immediately silenced.

I was recently watching a NOVA program about what scientists are doing to understand better what is causing the mega forest fires that the world has been experiencing for the past few decades. It was obvious that the problem was essentially manmade, a lack of understanding about what drives mega forest fires and the resulting bad forestry decisions that have exacerbated the problem significantly.

However, this was a PBS NOVA program and they couldn't leave it at that; they had to tell us that Climate Change (global warming) was a contributing cause that increased the danger significantly. So ... what did they suggest? Obviously, they wanted us to do all we could to minimize global warming.

This ended up being the focus ... let me remind you of a few facts:

- Every scientist and realist in the world believes in climate change.
- Every scientist and realist in the world believes that Climate Change is a natural process.
- Every scientist and realist in the world believes that Earth is currently warming.
- Every scientist and realist in the world believes that humanity is contributing to the warming of the planet.

Anyone telling you, or even suggesting that AGW "deniers" do not accept the above as facts, is manipulating you and you should not believe another word that comes out of their lying mouths.

So, what's the argument? Below are a number of issues of disagreement. I will add comments regarding the basic position of those that doubt the AGW hypothesis.

- *Has the AGW hypothesis been thoroughly tested and shown to be accurate in predictions based on the hypothesis?*

The decades-long pause in warming and the failure of the AGW models to accurately predict outcomes is proof that the hypothesis, as stated, is flawed; that there are other processes at work in Climate Change that we do not completely understand.

- *Have the proper scientific procedures been followed to determine the accuracy of the predictions made regarding the ultimate affect additional CO_2 will have on Earth's climate?*

The accepted scientific procedures for determining the validity of the AGW hypothesis were completely ignored. The rationale: there wasn't enough time, the end of the world was upon us!

- *Are we clear on the actual impact of humanity's activities on global warming? Do our actions have a significant impact or a very minor impact on the process?*

There is no clear understanding as to what impact humanity's actions are having on climate change. But, the amount of CO_2 we are adding to the atmosphere does not seem to be anywhere near enough to cause the predicted catastrophe that the public is being led to believe. Essentially the public is being manipulated so that those in power can achieve other ends that have nothing to do with climate change.

*****

I remember the screams of disaster when the BP oil leak pumped 200 million gallons of crude oil into the Gulf of Mexico. We were led to believe that the United States Gulf Coast was a complete disaster, with the media showing image after image of birds and fish covered with oil.

My wife and I toured the Gulf Coast during this time and we saw nothing but pristine white beaches empty of any sign of humanity. It was an economic disaster caused by the lying media and crooked politicians, not by the oil. In fact, when you consider the size of the Gulf of Mexico, the amount of oil from the BP leak amounted to a coke can of oil dumped into a football stadium full of water.

The media loves disasters, they create revenue and profits. Politicians have been schooled to never let a crisis go to waste. In the meantime, people suffer, but does the media care, do politicians care? No ... all that matters is profits and power.

- *What is the impact of additional CO_2 in the atmosphere? Is the increase in CO_2 positive or negative for Earth and the environment?*

There is strong research supporting the increase in CO_2 as a boon for life on Earth. CO_2 is an essential gas and there's strong proof that the plant kingdom is delighted with the change; that the planet is actually greening.[xlvii]

- *Do the models used to predict temperatures for the next century take into account all of the processes that have an impact on Earth's climate?*

The answer is obviously no, mostly because we don't know all of the processes involved in climate change. Climate scientists are just beginning to gain this essential understanding. However, there is one critical issue being completely overlooked, the impact of the Sun on climate change.

- *How is the climate affected by changes in the Sun's processes?*

The relationship of a "Quiet Sun" to Climate Change has been suspected for centuries. To ignore the research that shows this direct relationship is foolhardy to say the least and probably criminal, the type of activity that used to get propagators tarred and feathered and run out of town on a rail.

- *What will be the socioeconomic impact of policy changes in order to combat "global warming"?*

It appears that the impact would be catastrophic for the nation's socioeconomic health. In fact, even if we accept the truth of the AGW hypothesis, it could well be that the cure is actually much worse than the disease. Implementing the solutions being presented would have an immediate negative impact on the nation's socioeconomic health. Delaying such a massive solution might give us time to come up with one that is much more acceptable. This is especially true if the predictions are very wrong and the warming does not become anywhere near catastrophic.

- *Are current methods to combat global warming (solar cells, windmills, etc.) a net positive or negative to our socioeconomic health and/or the environment?*

It appears that the current methods are not efficient or effective.

- *Is the result obtained from the time, money, and resources spent to combat global warming worth the effort at any level?*

As you can see, there is no consensus on a number of Climate Change issues that are critical to making an intelligent decision on how we should move forward.

Regarding mega forest fires, with the knowledge that the planet is warming, instead of focusing on how to stop that from happening, we should be listening to the scientists who understand the issue and do what we can to limit the spread of mega forest fires.

Currently, there are actually laws in place that exacerbate the problem. For example, laws that limit the clearing of land. Just changing those laws and educating the public as to the need to keep the underbrush under control would do much to limit the number of mega forest fires that are occurring around the world.

For governments, progressives, the media, and the new "green" industry, their greatest fear is the AGW hypothesis being proven wrong. If this hypothesis fails, their world will come crashing down around them. Science will lose massive credibility with the result being a dramatic decrease in the funding received by individual researchers, research universities, and scientific organizations. It will take years, decades, for science to regain even a modicum of the respect that they have taken full advantage of during this time in our history.

The green industry will collapse, tens of trillions of dollars in investment will just disappear, millions will be out of work, and the world will suffer a major economic crisis. In addition, funding will dry up for some very important research, especially the research on new battery (energy storage) technologies. Environmentalists will be laughed at and organizations like Green Peace will lose funding, governments will lose their ability to overtax the masses. And, the effort to form one, powerful, and ultimately immoral, incompetent, ineffective world government will be set back decades. Essentially, all of these groups will discover that it wasn't heaven but hell that they had created.

Is there any doubt why all of these groups are so afraid of a Trump administration? He is the kink in their armor, the leak in their dam; he threatens the end of their world. They will do anything to destroy him and his administration, and I mean *anything*.

So, with literally trillions of dollars at risk and many special interest groups depending on the money, the Perfect Storm "scientific consensus" was born. There's a saying that a lie can make it halfway around the world before the truth gets its pants on.

In this instance the lie has the complete support of those scientists and scientific organizations whose funding and reputations depend on it; it has the support of the media who thrive financially and emotionally on every bit of bad news; it has the support of worldwide governments who see this as an excellent opportunity to gain power, influence, and the money needed to implement their progressive programs. And it has the support of a large portion of the entertainment industry that has its own reasons for wanting a global society. And, don't forget those that honestly believe that humans are the problem.

This time, the lie has a personal jet and the truth is left with a kiddie car.

Brad Fregger
June 2019

Addendum

"Extinction is the rule, survival is the exception."
                                          Carl Sagan

The Sixth Mass Extinction?

Science has indentified five nature-caused mass extinctions:

- End Ordovician, 444 million years ago, 86% of species lost
- Late Devonian, 375 million years ago, 75% of species lost
- End Permian, 251 million years ago, 96% of species lost
- End Triassic, 200 million years ago, 80% of species lost
- End Cretaceous, 66 million years ago, 76% of all species lost

In addition, many influential "scientific" organizations are very concerned about the coming "sixth mass extinction," the Holocene Extinction[xlviii] or the Anthropocene Extinction. This "mass extinction" is being blamed entirely on us and the disaster we are causing for other species as well as Earth ecosphere.

It is not only identified by the number of species that have become extinct because of our impact on Earth's environment but also by the decline in biodiversity and species populations.

This particular Sky Is Falling prediction has three main sources:

- The UN report by the Intergovernmental Science-Policy Platform on Biodiversity and Ecosystem Services (IPBES)
- World Wide Fund for Nature's (WWF) 2018 Living Planet Report
- 2015 research article "Accelerated modern human–induced species losses: Entering the sixth mass extinction" authored by Paul Ehrlich and numerous other contributors.

The IPBES issues presented are extremely easy to understand. Basically they are saying that extinctions levels are through the roof, more species going extinct that ever before in history. They state, categorically, that between 500,000 and 1 million plant and animal species face extinction. This includes 40 percent of amphibian species, more than 33 percent of all marine mammals and reef-forming corals, and at least 10 percent of insect species.

The WWF Living Planet Report isn't any more optimistic. Because of human activity and encroachment on animal habitat, species populations are declining at an unprecedented rate which is catastrophic to biodiversity and the precursor to mass extinction.

We shouldn't be surprised that the "research" article, "Accelerated modern human–induced species losses: Entering the sixth mass extinction" is the one that defines that current situation as the sixth mass extinction. After all, Paul Ehrlich is one of the major contributors and there's nothing he likes better than a catastrophic situation where humans are at fault.

So, once again, how did this happen and is it another major "scientific consensus"?

Like any "scientific consensus" or catastrophic hypothesis, the initial inspiration comes from an observation by concerned scientists. In this instance the biodiversity community noticed that species populations were declining at an accelerating rate. Further research proved this to be true and that there were essentially three major causes.

- The number of invasive species that were causing the populations of native species to decline and/or become extinct.
- Various forms of human activity that were making it very difficult for the species native to the habitat to survive.
- And, of course, global warming which was creating environments unsuitable for the native species.

The biodiversity scientists were, of course, very concerned. This was their bailiwick and the problem could easily upset the balance needed for a healthy environment, which they believed would cause many problems of the ecosystem and humanity.

Their concern was seen as an opportunity for Ehrlich, et al., "Never let a crisis go to waste." And they jumped on the bandwagon, doing their own research, which (of course) supported the initial hypothesis that human activity was playing a major role in the decline in species populations, which was having a negative impact on biodiversity; and then, extrapolating the research to support the contention that the Holocene Mass Extinction had truly begun.

But, Ehrlich was nowhere near the first to suggest that the Holocene or Anthropocene Mass Extinction was a potential danger to Earth and its inhabitants, animal and vegetable.

I'm not sure who was the first to suggest that human activity was causing extinctions at a rate which could be defined as a mass extinction but the following researchers, prior to Ehrlich, surely believed that this may well be happening:

In 1989, JM Diamond of the Department of Physiology, University of California Medical School, Los Angeles, wrote the research article, "The present, past, and future of human-caused extinctions."[xlix]

> Late-Pleistocene or Holocene extinctions of large mammals after the first arrival of humans in North America, South America and Australia may also have been caused by humans. Hence human-caused mass extinction is not a hypothesis for the future but an event that has been underway for thousands of years. As regards the future, consideration of the main mechanisms of human-caused extinctions (overhunting, effects of introduced species, habitat destruction, and secondary ripple effects) indicates that the rate of extinction is accelerating.

In 1996, Ben M. Waggoner wrote a small article for a University of California at Berkeley press, titled "The Holocene Epoch."[l]

> Humanity has greatly influenced the Holocene environment; while all organisms influence their environments to some degree, few have ever changed the globe as much, or as fast, as our species is doing. The vast majority of scientists agree that human activity is responsible for "global warming," an observed increase in mean global temperatures that is still going on. Habitat destruction, pollution, and other factors are causing an ongoing mass extinction of plant and animal species; according to some projections, 20% of all plant and animal species on Earth will be extinct within the next 25 years.

In June of 2001, Niles Eldredge wrote a research article for *Action Bioscience* titled, "The Sixth Extinction."[li] He believes that we are in a biodiversity crisis – "the fastest mass extinction in Earth's history …" In the article he stated that "We can stop the devastation of our planet and save our own species."

He cited five issues that needed to be handled if we were to save the world, they are:

- human destruction of ecosystems
- overexploitation of species and natural resources
- human overpopulation
- the spread of agriculture
- pollution

In 2014, Sacha Vignieri wrote the research article, "Vanishing fauna"[liii] (*Science*, July 2014) where he states,

> Although some debate persists, most of the evidence suggests that humans were responsible for extinction of this Pleistocene fauna, and *we continue to drive animal extinctions today*[21] through the destruction of wild lands, consumption of animals as a resource or a luxury, and persecution of species we see as threats or competitors.

As you can see, many researchers have postulated that we are in the throes of the sixth mass extinction, a mass extinction entirely the cause of humanity's encroachment into the natural habitat of thousands of species. We have been causing this problem ever since we have walked upon Earth.

To the average person and especially to those people who have a predisposition to the concept that humanity is destroying the planet, it appears that the argument is over, the proof is in ... and we are to blame.

Probably the best article written questioning the validity is, "Rethinking Extinction"[liii] written by Stewart Brand[liv] and published

---

[21] Italics mine.

in *Aeon*. Brand is the founder and editor of the *Whole Earth Catalog*. Brand was very concerned about the decline in wildlife populations but was very much against the hype of mass extinction as a selling point for that issue.

Below are a selection of quotes from the article.

> The way the public hears about conservation issues is nearly always in the mode of '[Beloved Animal] Threatened With Extinction'. That makes for electrifying headlines, but it misdirects concern.

> Viewing every conservation issue through the lens of extinction threat is simplistic and usually irrelevant. Worse, it introduces an emotional charge that makes the problem seem cosmic and overwhelming rather than local and solvable.

> Many now assume that we are in the midst of a human-caused 'Sixth Mass Extinction' to rival the one that killed off the dinosaurs 66 million years ago. But we're not.

> The fossil record shows that biodiversity in the world has been increasing dramatically for 200 million years and is likely to continue.

And finally,

> The phrase 'all currently threatened species' comes from the indispensable IUCN (International Union for Conservation of Nature), which maintains the Red List of endangered species. Its most recent report shows that of the 1.5 million identified species, and 76,199 studied by IUCN scientists, some 23,214 are deemed threatened with extinction. So, if *all* of those went extinct in the next few centuries, *and* the rate of extinction that

killed them kept right on for hundreds or thousands of years more, *then* we might be at the beginning of a human-caused Sixth Mass Extinction.

Hopefully, we aren't at the point where a whole lot of people are going to demand that we solve this problem, no matter what it takes, no matter who or what will suffer the consequences.

I want to know if there really is a problem. I've just presented a couple of examples of those who question the hypothesis, but there's still a conflict between those who support the hypothesis and those who don't.

In order to know for sure, we must fully understand the complexities that make up Earth's biosphere; a more complete understanding must be gained. Until then we cannot be sure as to the final impact of smaller populations of species nor even of the extinction of thousands, even millions of species.

In addition, we need to understand the following much better:

- Is there really a mass extinction in process, or even in our future?
- How many species have gone extinct in the past 500 years?
- How many species are there? Are we continually discovering new species?
- Have species identified as becoming extinct, resurfaced?
- Is it possible that a mass extinction could have a positive impact on Earth's ecological system?

And finally,

- Has the hypothesis,

    "The decline in species populations and the resulting lack of biodiversity has resulted in the extinction of many animal and vegetable species with the rate of extinction increasing to the point of where we are now in the midst of a human-caused mass extinction that will devastate Earth's biosphere and cause catastrophic problems for life on this planet."

    been fully tested?

In other words, "Are we confident as to the causes and results of the decline in species populations and the resulting impact on biodiversity and the extinction of species?"

Or, are we assuming the results because "they make sense" and "they fit our beliefs" as to the massive problems that are the result of human activities' devastation upon Earth; everything from overpopulation, encroachment on natural habitats, and catastrophic global warming.

Is There Really a Mass Extinction in Process?

Doug Erwin, a Smithsonian Paleontologist doesn't think so.[lv] In fact he's adamantly against the concept,

> It is absolutely critical to recognize that I am NOT claiming that humans haven't done great damage to marine and terrestrial [ecosystems], ... But I do think that as scientists we have a responsibility to be accurate about such comparisons.

Essentially, he agrees that the issue is worth looking into but that we need to be careful about how we present it to the public, that suggesting that we are in the midst of the sixth mass extinction is foolhardy, "junk science."

How Many Species Have Gone Extinct in the Past 500 Years?

Tech Times[lvi] suggests that the natural rate for species extinction is about 2 species to every 10,000 species every century. But, in the last 100 years over 500 species have become extinct. This data was the result of research done by a team of researchers led by UNAM[22] researcher Gerardo Ceballos; a fair estimate for the 500 years would be about 600 species, since, from their research there has been an increase is species extinction in the past 100 years.

The World Conservation Union, or IUCN, has documented the extinction of 849 species in the wild since 1500 A.D.

---

[22] "Autonomous National University of Mexico," UNAM) is a public research university in Mexico.

(when historical scientific records began). These include: birds, mammals, amphibians, and marine species.

These estimates are only guesses, "educated" guesses at best. We really have very little information as to the number of species that have become extinct.

How Many Species Are There?

Talk about a big unknown! The estimates range from an official figure of approximately 8.7 million[lvii] to a recent University of Arizona estimate of 2 billion.[lviii]

Smithsonian Magazine[lix] states that 300 new mammals are discovered every decade and they estimate that by 2050 the total known mammal species, now at about 5500, will increase to 7500. In addition, since 1500 only about 80 mammal species have gone extinct. The rate might be increasing but the increase is nowhere near the level suggested by the Sky Is Falling merchants of doom.

According to a Los Angeles Times article, 18,000 new species were discovered in 2016 alone.

There is no doubt we are just beginning to figure out how many species of plants and animals there are on Earth and how they relate to and support each other.

Have Species Identified as Being Extinct, Resurfaced?

CNN reported on an Australian research study[lx] that suggested that possibly as many as 1/3 of supposed extinct species are still alive. "Biologists at the University of Queensland examined more than 180 different extinct species, only to discover that a third of them were still alive." The researchers believed that conservationists have exaggerated the number of species that have been driven to extinction through human activity.

In addition, there are plenty of stories in the news and available through Google that talk about extinct species that have been found to still be alive.

In addition, some scientists are determined to bring back some extinct species through the use of modern technology. This attempt is very controversial with some concern as to how recreated species may impact the environment.

Is It Possible that a Mass Extinction Could Have a Positive Impact on Earth's Ecological System?

The truth be known, I've added this one just to get your attention. However, we do know that the result of past mass extinctions have resulted in a plethora of new species. In addition, we wouldn't exist if the comet hadn't hit Earth forcing the mass extinction of the dinosaurs. Bottom line, mass extinctions have, in the past, had a very positive impact on Earth's ecology.

Has the Mass Extinction Hypothesis Been Properly Tested?

The answer is simple, of course not! We don't have the data or knowledge necessary to properly test this hypothesis.

In a meeting I recently attended, of people who were convinced that a mass extinction was upon us, one of the leaders said, (paraphrase) "Catastrophic problems, demand catastrophic solutions." I'm sure he was making a play on Carl Sagan's famous phrase, "Extraordinary claims require extraordinary evidence."

In my opinion, the evidence we've been provided so far suggesting that we are in the midst of a human-caused mass extinction is nowhere near extraordinary, while the claim that we are the cause of the sixth mass extinction in history is definitely an extraordinary claim.

Conclusion

It seems to me that the two most influential organizations are actually lobbying strongly for a global initiative that would impact every human being on the planet and probably do very little to solve the essential problem of species population decline and biodiversity. However, the policies, rules, and laws created to "solve" this problem would have a significant impact on the socioeconomic health of our civilization, and that impact would not be a positive one for the vast majority of humanity.

Two major causes of species population decline are: Invasive species and changes in the environment of the natural habitat. Both have been going on since life appeared on this planet and there's nothing we can or should do to stop that natural process.

I'm not saying we shouldn't be good stewards of the environment and of the total ecosystem that makes up Earth, but we must be careful that the unintended consequences of our efforts do not cause more problems than they solve. We cannot do this right until we understand the entire process and the potential impact of any change we are attempting.

It appears that their hypothesis of massive species extinction is based on biased research, hopeful thinking, and a great need to save humanity by putting the elite in charge.

I'm on the side of Maharishi Mahesh Yogi, who said, "Control is the enemy of evolution." We don't need another governing body attempting to outdo nature while telling us how to live our lives.

It seems that IPBES is grasping at straws, at ways to convince the public that a catastrophe is in our future and they are the only ones with the solution. Sounds like the priests of ancient times to me. It also seems to be a rehash of the population bomb scenario put forth by Ehrlich.

In addition the WWF is presenting to us an updated version of the Polar Bear scenario that we now know was a total fraud, designed only to convince children and the childish of the danger, so that they would convince their elders.

Regarding Ehrlich, there's no doubt about his motivations; he's always made a mark by predicting the end of everything we know and love. This time is no different.

When will intelligent, caring people learn to discern what's important and what isn't. So many good people have been taken in by both catastrophic human-caused global warming and now, human-caused mass extinction. Where are the reasonable solutions to these issues? Where is the positive future in their predictions? While they might feel that saving the planet is the best positive future we can hope for; essentially, they are only hoping that their efforts will help and will not make matters worse, they don't have enough information to know otherwise. There is no doubt that their solutions will be carried on the back of the vast majority of humanity, while the elite continue to live in their white marble castles, wondering what we are complaining about and telling us to "eat cake."

It's time we began thinking out of the box, off of the planet. A spacefaring mentality would solve the issue of taking care of Earth's ecology. With millions, if not billions, living in space colonies, those left on Earth would treat it like a committed horticulturist treats their garden, with care, with love, and with total commitment to doing what's right.

Author's Bio

Figure 18: Brad Fregger

Brad Fregger, CEO and President of Groundbreaking Press, has 60 years combined experience in retailing, corporate training, publishing, software development, and teaching. He is an adjunct professor at Maharishi University of Management.

Brad is a practitioner/scholar, using the skills and knowledge he has learned to amass a remarkable record of accomplishment over the past 50 years. He's the founder of three major corporate training departments (Mervyn's, Atari, and Activision), and he's produced over 125 consumer, game, and business software products, published over 60 books and eBooks, and written 11 of his own.

Brad's amazing ability to complete projects on time and on budget, plus his creative management style, caught the attention of Tom Peters (*In Search of Excellence*), who then featured Brad in his book, *Liberation Management*.

Brad is an expert in many critical areas of business, from customer service, to negotiation, to technology management, and effective leadership. He presents talks, seminars, and workshops on a broad spectrum of subjects.

Brad holds a Master's Degree in Societal Futures. His speech, "*Earthward Implications of Cosmic Migration,*" was given at the American Astronautical Society's proceedings in honor of the tenth anniversary of Apollo 11's landing on the Moon.

As an adjunct professor at two Austin area universities (Texas State and St. Edwards), Brad developed and taught courses in the MBA program (Introduction to eCommerce, Human Relations) and the Master of Science in Organizational Leadership & Ethics (Leadership & Imagination) programs (Saint Edwards) and taught Business Management and Business Communications (Texas State).

Brad is currently Director of Career Services for the MBA program at MUM (Maharishi University of Management) in Fairfield, Iowa, where he resides with his wife Barbara.

Contact

Groundbreaking Productions coordinates all of Brad Fregger's speaking engagements and offers seminars and workshops based on the concepts presented in Brad's books. (www.goundbreaking.com).

Are You an Author Looking for a Publisher?

Groundbreaking Press assures its author-clients a quality product that meets the highest standards of the publishing industry.

For additional information, see:
www.groundbreaking.com

Purchasing Multiple Copies

Many organizations have requested multiple-copy pricing of *Sky Is Falling Myths*. We offer the following:

5-10 copies: 15% discount
11-25 copies: 25% discount
26-50 copies: 35% discount
51 copies or more: 50% discount

Contact us to take advantage of this pricing:

Phone: 512-657-8780
Email: brad@groundbreaking.com
Alternate eMail: bradfregger@gmail.com

End Notes

[i] https://en.wikipedia.org/wiki/Giordano_Bruno
[ii] https://www.investors.com/politics/commentary/the-global-warming-thought-police-want-skeptics-in-jail/
[iii] https://en.wikipedia.org/wiki/Earth_goddess
[iv] http://www.worldhunger.org/2015-world-hunger-and-poverty-facts-and-statistics/#progress
[v] https://en.wikipedia.org/wiki/List_of_dates_predicted_for_apocalyptic_events
[vi] http://www.forbes.com/sites/henrymiller/2012/09/05/rachel-carsons-deadly-fantasies/#111175d715d1
[vii] https://www.acsh.org/news/2016/02/11/how-poisonous-is-ddt
[viii] http://www.aei.org/publication/18-spectacularly-wrong-apocalyptic-predictions-made-around-the-time-of-the-first-earth-day-in-1970-expect-more-this-year-3/
[ix] https://www.theguardian.com/lifeandstyle/2011/jul/17/food-prices-rise-commodities
[x] https://en.wikipedia.org/wiki/Coal_power_in_the_United_States
[xi] http://www.demog.berkeley.edu/~andrew/1918/figure2.html
[xii] https://www.bbc.com/news/world-us-canada-44416727
[xiii] https://www.lenntech.com/greenhouse-effect/global-warming-history.htm
[xiv] https://en.wikipedia.org/wiki/Hockey_stick_controversy
[xv] https://en.wikipedia.org/wiki/Nicolaus_Copernicus
[xvi] http://www.inquiriesjournal.com/articles/1675/copernicus-galileo-and-the-church-science-in-a-religious-world
[xvii] https://www.chronicle.com/blogs/innovations/failed-hypotheses-in-academe-and-beyond/28980
[xviii] https://phys.org/news/2018-07-gault-site-date-earliest-north.html
[xix] https://www.scientificamerican.com/article/ancient-bones-spark-fresh-debate-over-first-humans-in-the-americas/
[xx] https://www.tandfonline.com/doi/abs/10.1080/17512433.2018.1519391
[xxi] https://www.telegraph.co.uk/science/2016/06/12/high-cholesterol-does-not-cause-heart-disease-new-research-finds/
[xxii] https://www.ipcc.ch/about/history/
[xxiii] https://www.sciencedaily.com/releases/2019/04/190426075447.htm
[xxiv] https://www.thenewamerican.com/tech/environment/item/18888-embarrassing-predictions-haunt-the-global-warming-industry
[xxv] https://www.climatechangenews.com/2018/06/05/green-economy-now-worth-much-fossil-fuel-sector/
[xxvi] https://www.ncbi.nlm.nih.gov/pmc/articles/PMC1420798/
[xxvii] https://www.nobelprize.org/prizes/physics/2011/summary/
[xxviii] https://phys.org/news/2016-10-Universe-rateor.html
[xxix] https://www.cnsnews.com/news/article/barbara-hollingsworth/4-peer-reviewed-studies-find-no-observable-sea-level-effect-man
[xxx] https://www.sciencedirect.com/science/article/pii/S2213305415300199

[xxxi] https://www.westernjournal.com/scientist-medias-misinformation-global-warming-wildfires/
[xxxii] https://www.investors.com/politics/editorials/despite-what-youve-heard-global-warming-isnt-making-weather-more-extreme/
[xxxiii] https://www.thegwpf.org/as-polar-bear-numbers-increase-gwpf-calls-for-re-assessment-of-endangered-species-status/
[xxxiv] https://youtu.be/Pxq4PmFV6yg
[xxxv] https://www.iisd.org/pdf/2008/climate_forced_migration.pdf
[xxxvi] https://www.nasa.gov/feature/goddard/2016/carbon-dioxide-fertilization-greening-earth
[xxxvii] https://academic.oup.com/icesjms/article/73/3/529/2459146
[xxxviii] https://wattsupwiththat.com/2018/06/05/the-total-myth-of-ocean-acidification/
[xxxix] https://arizonadailyindependent.com/2014/01/28/the-myth-of-ocean-acidification-by-carbon-dioxide/
[xl] https://www.sciencealert.com/reefs-may-benefit-from-global-warming
[xli] https://www.lomborg.com/
[xlii] http://joannenova.com.au/2011/06/killing-people-with-concern-biofuels-lead-to-nearly-200000-deaths-est-in-2010/
[xliii] https://atlasmonitor.wordpress.com/2016/08/10/quiet-sun-may-cause-global-cooling-says-newcastle-astrophysicist/
[xliv] http://www.nature.com/nclimate/journal/v6/n8/full/nclimate3004.html
[xlv] http://www.co2science.org/education/book/2011/55BenefitsofCO2Pamphlet.pdf
[xlvi] https://phys.org/news/2016-10-Universe-rateor.html
[xlvii] https://www.nasa.gov/feature/goddard/2016/carbon-dioxide-fertilization-greening-earth
[xlviii] https://en.wikipedia.org/wiki/Holocene_extinction
[xlix] https://www.ncbi.nlm.nih.gov/pubmed/2574887
[l] https://ucmp.berkeley.edu/quaternary/holocene.php
[li] http://www.actionbioscience.org/evolution/eldredge2.html
[lii] https://science.sciencemag.org/content/345/6195/392
[liii] http://markkit.net/untrusted/aeon.co_magazine_science_why-extinction-is-not-the-problem_.html.gz
[liv] https://en.wikipedia.org/wiki/Stewart_Brand
[lv] https://www.theatlantic.com/science/archive/2017/06/the-ends-of-the-world/529545/
[lvi] https://www.techtimes.com/articles/64542/20150630/humans-cause-of-extinction-of-nearly-500-species-since-1900.htm
[lvii] https://www.bbc.com/news/science-environment-14616161
[lviii] https://phys.org/news/2017-08-biodiversity-earth.html
[lix] https://www.smithsonianmag.com/science-nature/meet-the-new-species-748819/
[lx] http://www.cnn.com/2010/WORLD/asiapcf/09/29/extinct.species.animals/index.html

www.ingramcontent.com/pod-product-compliance
Lightning Source LLC
Chambersburg PA
CBHW071409220526

45469CB00004B/1215